T0180999

The Lean Product Design and Development Journey

Marcus Vinicius Pereira Pessôa
Luís Gonzaga Trabasso

The Lean Product Design and Development Journey

A Practical View

 Springer

Marcus Vinicius Pereira Pessôa
Instituto Tecnológico de Aeronáutica
São José dos Campos, São Paulo
Brazil

Luís Gonzaga Trabasso
Instituto Tecnológico de Aeronáutica
São José dos Campos, São Paulo
Brazil

Additional material to this book can be downloaded from http://extras.springer.com.

ISBN 978-3-319-83586-0   ISBN 978-3-319-46792-4   (eBook)
DOI 10.1007/978-3-319-46792-4

Printed on acid-free paper

This Springer imprint is published by Springer Nature
The registered company is Springer International Publishing AG
The registered company address is: Gewerbestrasse 11, 6330 Cham, Switzerland

To my wife Liane, my son João Pedro and my daughter Anna Clara, without whom this work would not be possible.

—Marcus Vinicius Pereira Pessôa

To my wife Rachel, my sons Rafael, Gabriel, Tomás and my daughter Júlia who inspire my daily life.

—Luís Gonzaga Trabasso

Preface

Product design and development can be understood as some kind of information-based factory performing the product development process (PDP). The goal of the PDP was to create a concept or design for producing a product, which reduces risk and uncertainty while gradually developing a new and error-free product, which can then be realized by manufacturing, selling, and delivering to the customer. PDP itself is people-based, complex, and nonlinear, with high ambiguity and uncertainty. Consequently, a wide spectrum of variables can affect its success, and not surprisingly, over time, over budget, and low quality are commonplaces on PDP.

Through this book, we aim to present a series of high-performance product design and development best practices that can support creating or improving a product development organization. Rather than being a book about Toyota or any other company applying lean, this book is strongly rooted in the lean philosophy and includes discussions of systems engineering, design for X (DFX), agile development, integrated product development, and project management.

The "Lean Journey" proposed herein takes a value-centric approach, where the lean principles application to PDP let the choice of tools and methods emerge from the observation of the particularities of each company. Therefore, learning lean product development (LPD) is not about learning tools, but rather understanding how to apply the philosophy. Indeed, the lean journey is about mind-set and culture change rather than adopting tools and techniques. Many of the tools and techniques already in use in your company might be used in the lean way.

The scope of the book includes university students majoring in engineering and professionals working in the various fields of engineering as well as in related fields outside of engineering.

We have been using the book's contents to teach a "lean product development" course to graduate students in engineering for the past six years. Based on this success, we intend to reach Product development and lean product development university courses at the graduate and undergraduate levels as well.

Since the design and development of products is the aim of most engineering areas, we target engineering courses in a broad sense, both undergraduate and graduate.

Also, the growing trend of lean makes the book suitable to summer courses and short courses that aim practitioners and people interested in applying the lean product development techniques in their daily use. In this case, we can reach a broader audience than engineers since product development is a multidisciplinary endeavor. Particularly, we have a close relation with business and marketing professionals, since they also deal with introducing the right product into the market.

Indeed, during the past six years, we taught people with backgrounds in engineering (mechanical, electrical, mechatronics, software, and chemical) and those in related fields such as business, logistics, industrial design, Web site design, and law.

The light, straightforward, and practical narrative makes the book suitable to be read by companies' executives that are interested in better understanding how the lean philosophy suits the development of products, services, and products as services in their particular companies.

We strongly believe that the book's practical approach is accessible to this broad audience.

We would like to thank some people without whom this book would not be feasible: the support from the Brazilian Air Force, particularly from Ricardo Ferreira Gomes dos Santos (Colonel) and Carlos Vuyk de Aquino (Brigadier), who believed that knowledge should be shared, thus allowing Marcus dedicating some of his working time to the book writing; the good discussions about product development, systems engineering, and lean topics with Prof. Warren Seering (MIT), Dr. Eric Rebentisch (MIT), Dr. Geilson Loureiro (INPE), and Dr. Juan Jauregui Becker (University of Twente); the very interesting and live conversation on integration concepts, which were held with Prof. J.R. Hewit (Loughborough University) and Prof. M.M. Andreasen (Technical University of Denmark); and ITA's lean product development course attendees (classes 2011–2016), once through the interaction with them the book was shaped. Particularly, from class 2015, we would like to thank Priscila Malaguti Guerzoni who co-authored Chap. 14 and Wesley Rodrigues de Oliveira and André Vinicius Santos Silva, who co-authored Chap. 15. We would like to thank all the researchers of the Competence Center of Manufacturing at ITA (CCM/ITA), a laboratory that hosts strategic projects with industrial partners and that applies successfully the concepts presented herein. Last but not least, a special thanks to Marcus' children João Pedro and Anna Clara, who prepared Figs. 2.18, 7.5, and 7.6 and Figs. 3.7, 4.6, and 7.2, respectively.

Finally, we acknowledge the work from all the cited authors whose research created the necessary foundation to the work presented here. We highlight that every effort was made to find holders of copyrights and publishers will be happy to correct errors or omissions in future editions.

São Paulo, Brazil   Marcus Vinicius Pereira Pessôa
                                                                          Luís Gonzaga Trabasso

Contents

Abbreviations and Acronyms

3D	Tridimensional
AC	Actual Cost
ACWP	Actual Cost of Work Performed
BCWP	Budgeted Cost of Work Performed
BCWS	Budgeted Cost of Work Scheduled
Brep	Boundary representation
CAD	Computer-aided design
CBA	Cost–benefit analysis
CDR	Critical design review
CE	Chief engineer
CFR	Code of Federal Regulations
CG	Center of gravity
CIM	Computer integrated manufacturing
CPI	Cost performance index
CSG	Constructive solid geometry
CV	Cost variance
DbF	Design by features
DFA	Design for assembly
DFC	Design for correction
DFEB	Design for E-Business
DFM	Design for manufacturability
DFMA™	Design for manufacture and assembly
DFMt	Design for maintainability
DFP	Design for packing
DFR	Design for recycling
DFRel	Design for reliability
DFSft	Design for safety
DFSv	Design for service
DFX	Design for excellence
DMAIC	Define, measure, analyze, improve, and control
DTC	Design to cost

DTCG	Design to center of gravity
DTNP	Design to net power
DTW	Design to weight
DTX	Design to excellence
ECF	Estimate cost of a function
EDS	Engineering development system
EV	Earned value
EVA	Earned value analysis
FAA	Federal Aviation Administration
FAPESP	São Paulo State Research Support Foundation
FFS	Full flight simulator
FHFT	Flight hardware functional test
FMT	Functional model test
FQR	Final qualification review
GT	Ground test
HR	Human resources
IBGE	Brazilian Institute of Geography and Statistics
IDV	Integrative design variables
IMVP	International Motor Vehicle Program
IPD	Integrated product development
IQ	Information quality
IT	Information technology
ITA	Aeronautics Institute of Technology
JIT	Just-in-time
JIT-DM	Just-in-time decision making
KBE	Knowledge-based engineering
KM	Knowledge management
KSA	Knowledge, skills, and attitudes
LPD	Lean product development
LPDD	Lean product design and development
LPDO	Lean product development organization
MDT	Module development teams
MIT	Massachusetts Institute of Technology
MoE	Measure of effectiveness
MTBF	Mean time between failures
MTBM	Mean time between maintenance
OLED	Organic light-emitting diode
PD	Product development
PDCA	Plan, Do, Control, Act
PDP	Product development process
PDR	Preliminary design review
PDS	Product development system
PDVMB	Product development visual management boards
PMBOK	Project management body of knowledge

PMI	Project management institute
PR	Perceived risk
PT	Tailchute's parachute subsystem test
PV	Planned value
QFD	Quality function deployment
R&D	Research and development
RACI	Role & responsibility charting
SBCE	Set-based concurrent engineering
SDR	System design review
SIDS	Sudden infant death syndrome
SIVOR	Flight simulator with robotic motion platform
SOW	Statement of work
SPD	Serial product development
SPI	Schedule performance index
SV	Schedule variance
TCF	Target cost of a function
TPS	Toyota production system
TV	Total value
VE	Value engineering
VFD	Value function deployment
VSMA	Value stream mapping and analysis
WIP	Work in process

Introduction

Product design and development can be understood as some kind of information based factory performing the product development process (PDP). The goal of the PDP is to create a concept or design for producing a product, which reduces risk and uncertainty while gradually developing a new and error-free product, which can then be realized by manufacturing, selling, and delivering to the customer.

PDP is a problem-solving and knowledge-accumulation process, which is based on two pillars: "do the thing right" and "do the right thing." The former guarantees that progress is made and value is added by creating useful information that reduces uncertainty and/or ambiguity [1, 2]. The latter addresses the challenge to produce information at the right time, when it will be most useful [3, 4]. Developing complex and/or novel products and systems multiplies these challenges; the coupling of individual components or modules may turn engineering changes in a component into "snowballs," in some cases causing long rework cycles and making it virtually impossible to anticipate the final outcome [5]. Not surprisingly, over time, over budget, and low quality are commonplace on PD projects.

PDP evolved from artisanal, highly customized product development prior to the industrial revolution in the eighteenth century. This period brought serial production onto the scene, extending its principles to the PDP as well. At that time, the PDP was characterized by highly specialized personnel, low communication among company departments, and standardized products. This product development process is referred as *Serial* or *Sequential PDP* [6–8]. A number of problems emerged from the lack of integration within the Serial PDP, such as fragmented views of the same product, and not taking into account production issues until the PDP late stages, when product modifications, if needed, are more difficult and costly to carry out than those in the PDP early stages.

To overcome these problems, an integrated approach to PDP emerged in early 1990s, known by a myriad of names: *Concurrent Engineering, Simultaneous Engineering, Integrated Product Development, New Product Development*, and so forth [9]. Regardless the name, these approaches aim to rescue the integration aspect that was lost by the Serial PDP; the product and production planning processes are then conduced in an integrated and simultaneous manner. The integration proposal was constantly enlarged to encompass the whole product life cycle, starting from the customer needs and including the conceptual and detailed design, production, use, and discard.

A step further from the integrated approach was taken by the Toyota Motor Company, where the company-wide *lean philosophy* application allowed the emergence of a PDP that has consistently succeeded in its product development projects, presenting productivity better than their rivals [10, 11].

To deliver better products faster and cheaper, some firms are attempting to use the same principles as Toyota's and create *lean development* processes that continuously add customer value (i.e., that sustain a level of "progress" toward their goals) [12, 13]. This movement toward "lean" is not limited to physical product development companies, but also to systems, services, and information development. In order to succeed in this endeavor, duplicating some lean tools and techniques does not suffice, and the company has to understand the principles behind them and how these principles apply to the company's culture [14].

The Book's "Lean Journey"

Through this book, we aim to present a series of high-performance product design and development best practices that can support creating or improving a product development organization. Rather than being a book about Toyota or any other company applying lean, this book is strongly rooted in the lean philosophy and includes discussions of systems engineering, design for X (DFX), agile development, integrated product development, and project management.

One of the first challenges we faced while writing this book was to embed the lean philosophy into it. Lean thinking (or philosophy) is a way to specify value, align the value-added actions, execute these actions without interruption, and improve continuously.

In PD, adding customer value can be less a function of doing the right activities (or of not doing the wrong ones) than of getting the right information in the right place at the right time. Hence, the focus of lean must not be restricted to activity "liposuction" (waste reduction), but must address the PD process as a system (value creation).

The "Lean Journey" proposed here takes a value-centric approach, where the lean principles application to PD let the choice of tools and methods emerge from the observation of the particularities of each company. Therefore, learning lean product development (LPD) is not about learning tools, but understanding how to apply the philosophy.

In fact, many of Toyota's tools, techniques, and practices are countermeasures developed according to its necessity and capacity and are fit to a particular situation and moment due to the restrictions that are in place. Therefore, instead of studying the solutions developed to solve previous problems bounded by its environment and particularities, we should learn how to go about developing those solutions. Once the future lies beyond the horizon, today's solutions may not continue to be effective. In reality, some of the actual problems may be the result of some previously applied solutions. The competitive advantage lies on the ability to understand conditions, then creating and fitting smart solutions. Focusing only on solutions does not make an organization adaptive and does not create a sustainable competitive advantage.

The Lean Wheel

In order to deliver the proposed value, the book was structured using the metaphor of a "Lean Wheel System" (Fig. 1). The wheel shows pictorially that the tools, techniques, and processes are means and not the end; the lean philosophy itself and the concepts of making value and reducing waste are at the core of the LPD system.

The Lean Wheel System is composed of the following elements:

- **The Track**: Each wheel has to be designed considering the terrain where it will be used; in this case, the environment is composed of the product development characteristics, particularly its relation to the market the company is in. Therefore, the concepts, tools, and techniques presented in this book might not apply to different "tracks."

Fig. 1 The lean wheel system

- **The Car**: Any wheel has to fit the car it is attached to. Failing to mount on the car will reduce its capacity to provide high performance and a safe drive. In this metaphor, the car represents the whole company for which a successful stream of new products moves it forward.
- **The Wheel Hub**: The hub guarantees that the lean product development initiatives are not alone, but connect to the whole lean enterprise (the car). Included in this part is the "Core Lean," composed of continuous improvement and the concepts of value and waste which are applied to the product development.
- **The Wheel**: The wheel itself includes all supporting organizational aspects encompassing the lean product development organization (LPDO) culture: organizational structure, knowledge management, and continuous improvement aspects.
- **The Tire**: This is the part that actually interfaces with the track and includes Lean PDP, Lean PD Tools, and Lean PD Techniques.

The closer you get from the wheel hub, the more general are the discussed concepts. Value, waste, and continuous improvement are concepts that permeate the lean philosophy application to any domain or type of process. The wheel elements, while are also very general in the essence, were somehow shaped in a way to provide the interface with the lean core elements and the PDP. By changing from the PDP to any other of the company's processes, you might expect some adaptations in the wheel. Finally, the tire elements are fitted to the process (track) they will support.

Make Sure You Adapted the Wheel to Your Car

Although this book uses the Toyota Motor Company as a reference to the lean product development organization, you must be warned that:

1. The examples show how Toyota worked at a particular moment in time, so the company may have evolved in several ways as the years passed. For the sake of the examples and the presentation of ideas and attitudes, they remain valid.
2. Toyota is a final integrator of complex automobile products aimed at global markets. The reader's company may target quite different markets, with very different products, requiring adjustments of the presented concepts.

The same remark is valid to any other examples presented in the text. You must remember that any company can copy techniques and practices, or purchase the tools and technology used by any other company. Successful utilization of such techniques, practices, tools, and technology, though, depends on the ability to customize them in a way that makes them fit to the unique reality of the company using them. Toyota has had the foresight and discipline to customize a high-performance product development system to fit within a broader framework, one that includes all the processes—from design to manufacture—within the company.

Indeed, the lean journey is about mind-set and culture change rather than adopting tools and techniques. Many of the tools and techniques already in use in your company might be used in the lean way.

Our experience, though, shows that it is very difficult for a person used to applying tools and techniques with the mind-set bounded by a certain paradigm to do that in a different way. Unconsciously, he or she turns back into the previous way. This is the reason we proposed the value function deployment (VFD) technique. The VFD acts as a backbone of the lean product development process, always reminding the practitioner about the lean directives while he/she can apply the tools and techniques he is accustomed to.

Book Structure

The book is structured according to the Lean Wheel System and consists of five parts (Fig. 2).

We considered discussions about the company as a whole (the car) outside the book's scope. Therefore, the focus is maintained on the part of the company in charge of product development, which we aim to turn into a lean product development organization (LPDO).

Part I sets the track on which the Lean Wheel System was designed to roll. It discusses the general concepts and characteristics from the product development system. It also presents the evolution of the product development process from artisanal, integrated, and finally lean product development.

Part II describes the lean thinking's core elements. According to lean thinking, all the company's efforts should be on delivering value, while anything different from that is considered waste. Once eliminating all the waste is proven rather difficult, if not impossible, continuous improvement keeps the companies' processes from slipping backward and moves toward a desired state which delivers full value.

In Part III, the discussion focuses on the cultural and knowledge management aspects that an LPDO should consider to support the actual embedding of the lean philosophy into the PDP. Even though there is no chapter dedicated to the organizational structure itself, aspects related to that are presented in Chaps. 7 and 8.

Part IV goes deep into the lean product development process and dedicates separate chapters to describe each LPD phase in detail.

Finally, Part V gives some "on the road examples" and discusses some bumps one might expect while on the track of one's LPD journey.

All chapters have a "Practical View" section where we add value, including insights from our experience while applying these very concepts.

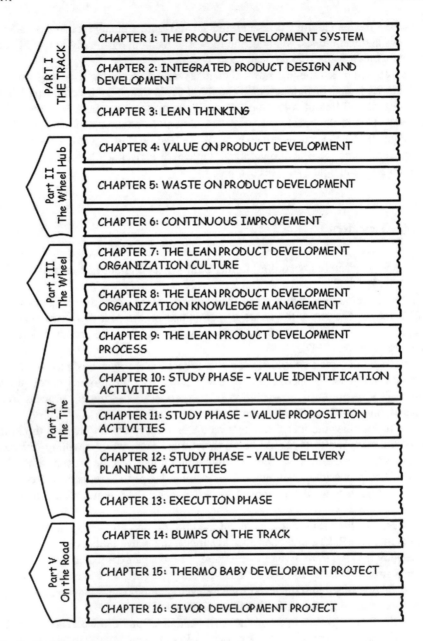

Fig. 2 The book's structure

Compare this journey to learning to fly aircraft. You have to understand the meteorology and how to navigate; this knowledge of the environment (the track) is paramount. Without it, you will not understand how the environment might affect you. In the same way, understanding the product development process and how it evolved helps you to perceive changes in the environment and how it is going to evolve.

In the same example, the domain of the wheel hub elements is as important knowing aerodynamics; when flying a machine, one must understand the physics of flight; when designing and developing in the lean way, one must understand the elements that support the lean thinking emergence.

The wheel hub compares to the aircraft subsystems (motor, hydraulic, electrical, instruments, etc.), which you study in order to understand how the machine works and how to command it.

In order to guarantee the safe flight, procedures are defined (the tire) to each flight stage, from preflight to post-flight, going through takeoff, cruise, landing, etc. This includes both normal and emergency procedures.

Finally, we learn from experienced pilots how to link everything together (all the wheel's elements) in order to avoid accidents and incidents.

In the same way, we suggest the reading (or teaching) this book. The chapter sequence supports a knowledge journey where what you learned from the previous chapters supports your further progress.

References

1. De Meyer A, Loch CH, Pich MT (2002) Managing project uncertainty: from variation to chaos. Sloan Manage Rev 43(2):60–67
2. Schrader S, Riggs WM, Smith RP (1993) Choice over uncertainty and ambiguity in technical problem solving. J Eng Technol Manage 10:73–99
3. Murman et al (2002) Lean enterprise value: insights from MIT's lean aerospace initiative. Palgrave, New York
4. Thomke S, Bell DE (2001) Sequential testing in product development. Manage Sci 47(2):308–323
5. Mihm J, Loch CH, Huchzermeier A (2002) Modeling the problem solving dynamics in complex engineering projects. INSEAD working paper. INSEAD, Fontainebleau, France, March 2002
6. Pahl G, Beitz W, Feldhusen J, Grote KH (2007) Engineering design: a systematic approach, 3rd edn. Springer, London
7. Huang GC (1996) Design for X: concurrent engineering imperatives. Chapman & Hall, London
8. Krishnan V, Ulrich K (2001) Product development decisions: a review of the literature. Manage Sci 47(1):1–21
9. Andreasen MM, Hein L (1987) Integrated product development. IFS-Springer Verlag, Berlin
10. Kennedy MN (2003) Product development for the lean enterprise. Oaklea Press, Richmond
11. Womack JP, Jones DT, Ross D (1990) The machine that changed the world. Rawson Associates, New York
12. Browning T, Deyst J, Eppinger S (2002) Adding value in product development by creating information and reducing risk. IEEE Trans Eng Manage 49(4):443–458
13. Walton M (1999) Strategies for lean product development: a compilation of lean aerospace initiative research. Research paper 99-02, Massachusetts Institute of Technology, Cambridge
14. Rother M (2010) Toyota Kata: managing people for improvement, adaptiveness and superior results. McGraw-Hill, New York

About the Authors

Marcus Vinicius Pereira Pessôa, D.Sc. PMP, is a graduated military pilot from the Brazilian Air Force Academy and has master of science in applied computing from Instituto Nacional de Pesquisas Espaciais, Brazil (1998), and D.Sc. in mechanical engineering—Instituto Tecnológico de Aeronáutica, Brazil (2006). He was also a postdoctorate fellow at MIT mechanical and systems engineering, USA (2007–2008). In the Brazilian Air Force, he worked on systems development, systems acquisition, and development process improvement. Currently, he is a consultant on the lean product development and project management areas and teaches the lean product development course at the Instituto Tecnológico de Aeronáutica, Brazil.

Luís Gonzaga Trabasso, Ph.D., is a graduated mechanical engineer from Universidade Estadual Paulista Júlio de Mesquita Filho, Brazil (1982) and has master of science in engineering and aerospacial technology from Instituto Nacional de Pesquisas Espaciais, Brazil (1985) and Ph.D. in mechanical engineering—Loughborough University, England (1991). He has held various positions as professor at Instituto Tecnológico de Aeronáutica, Brazil (ITA) since he entered in 1984. He is one of the founders of the Competence Center of Manufacturing at ITA (CCM/ITA), a laboratory that hosts strategic projects with industrial partners such as EMBRAER, SIEMENS, FIAT, REXAM, and PETROBRAS. Currently, he is full professor at ITA, focusing his research on integrated product development (IPD), Lean IPD, industrial automation, and robotics.

Part I
The Track

Part I defines the "environment" that gives this book context, by presenting the evolution of product design and development approaches (Fig. 1). Chapter 1 shows the particularities of the Product Development System, with special emphasis on the Product Development Process. Chapter 2 discusses the evolution of product development approaches through time. Finally, Chap. 3 presents the lean philosophy and its implications into the Product Development Process.

People already familiarized with these subjects might skip this part. We suggest, though, that you invest some time in reading the concepts presented in the gray boxes, and the "A Practical View" sections included at the end of each chapter.

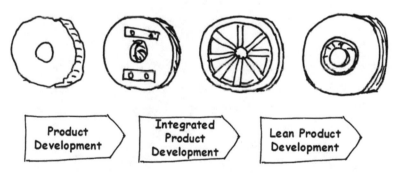

Fig. 1 Product design and development approaches evolution

Chapter 1
The Product Development System

This chapter shows the particularities of the Product Development System, with special emphasis to the Product Development Process (PDP). PDP itself is people-based, complex, and non-linear, with high ambiguity and uncertainty. Consequently, a wide spectrum of variables can affect its success, and, not surprisingly, over time, over budget and low quality are commonplaces on PD projects. By discussing the PDP characteristics and its consequences, we aim to show that having a high performance PDP is not an easy task to any company; therefore competitive advantage comes from accepting these particularities and understanding how they affect your particular PDP. Far from neglecting these particularities, the lean company deeply understands them, how they affect its particular reality, and shape its PDP to exploit its strengths and avoid its weaknesses.

1.1 Introduction

The Product Development System (PDS) is an organizational system that manages both the product portfolio and each individual product development. A high performance PDS, therefore, is capable of consistently articulating market opportunities that match the enterprise's competencies and executing the Product Development Process (PDP), thereby guaranteeing that progress is made and value is added by creating timely results [1].

The PDS, thus, is the interface between the enterprise and the market, being responsible for the identification, and even the anticipation, of the market's needs in order to propose solutions to fulfill those needs [2, 3].

According to the General Systems Theory [4, 5], the PDS falls in the category of *open systems*, since it has the characteristic of influencing and being influenced by the environment (as opposed to *closed systems*, which do not allow feedback). As any system, the PDS is composed of (Fig. 1.1): (1) inputs—the material, energy, or information that enters through the boundaries of the system; (2) outputs—the material, energy, or information that passes through the boundaries of the system; (3) process or throughput—the process of conversion or

© Springer International Publishing AG 2017
M.V.P. Pessôa and L.G. Trabasso, *The Lean Product Design and Development Journey*, DOI 10.1007/978-3-319-46792-4_1

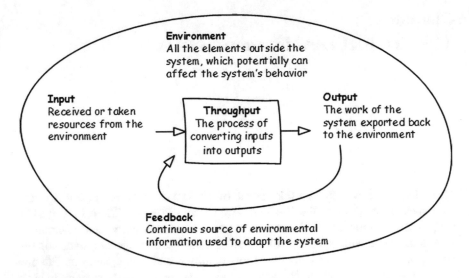

Fig. 1.1 Generic system's elements

transformation of inputs into outputs; and (4) the environment that is outside the boundaries of the system.

Every system performs a purposeful action, which is the function, and each element of the system interacts at least with another one: the PDS purpose is performing the Product Development Process (PDP). Through the PDP, the information is turned into specifications, or some sort of "product recipe," to be produced. Ulrich and Eppinger [3] define Product Development (Process) as the set of activities from the market opportunity perception to the production, sale, and delivery of a product.

To illustrate, Fig. 1.2 presents some PDP models found in the literature, and how their scopes relate to the market life cycle of a product [6]. This cycle includes all stages from the product conception until its discontinuity, while the enterprise works to make and keep the product competitive.

Development Stage—Comprises the PDP activities, from the identification of the market's needs, concept development and product and process engineering that end with a product, a process, and any mix of products and processes that can be delivered, sold or produced. During this stage Integrated Product Development (IPD), Systems Engineering (SE), and Project Management (PM) play important roles.

Introduction Stage—This stage of the cycle is normally the most expensive for a company launching a new product. The size of the market for the product is small, which means sales are low, until you increase the market. On the other hand, the cost of research and development, consumer testing, and the marketing needed to launch the product can be very high, particularly considering a

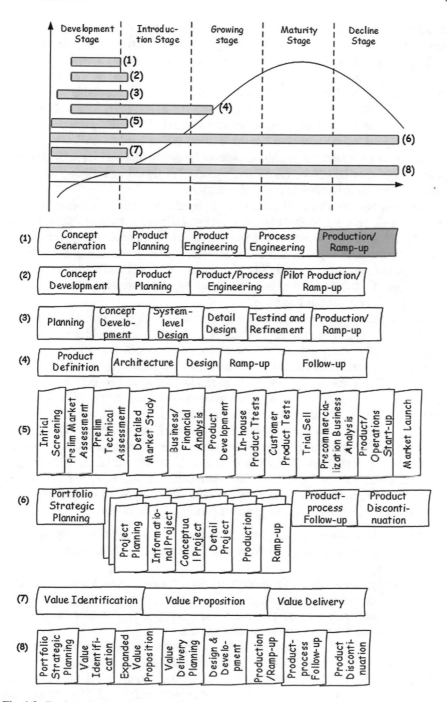

Fig. 1.2 Product development process models

competitive sector. Successful products are the ones that capture the aspects valued by early adopters, and that give strong support to marketing communications seeking to build awareness and to educate potential consumers about the product.

Growth Stage—The growth stage is typically characterized by a strong growth in sales and profits. The company can start to benefit from economies of scale in production, increasing the profit margins, and the total profit. As a result, more money is invested in promotional activities, maximizing the potential of this growth stage. Competition also begins to increase which in turn leads to price decreasing. As a strategy to maintain product quality, additional features and support services may be added. Therefore, a product designed considering the whole value chain is more flexible to these adaptations.

Maturity Stage—During the maturity stage, the product is established and the company's objective is maintaining the market share it has built up. This is probably the most competitive time for most products and the company must invest wisely in any marketing they undertake. Product modifications or improvements to the production process, which might give some competitive advantage, shall also be considered. Modular, design for manufacturing, and assembly products give the company advantage at this stage.

Decline Stage—Eventually, the market for a product will start to shrink, and this is what's known as the decline stage. This shrinkage could be due to the market becoming saturated (i.e. all the customers who will buy the product have already purchased it), or because the consumers are switching to a different type of product. While this decline may be inevitable, it may still be possible for companies to make some profit by switching to less-expensive production methods and cheaper markets, or finding new uses for the product.

In order to allow for comparative analysis, the PDP models in Fig. 1.2 were represented: (1) as sequential processes, even though they might have several cycles, parallel tracks, and fuzzy frontiers; and (2) on the initial life cycle stages, although additional development can be made later as a way to evolve the product or fix problems, adapting it to new requirements and postponing the end of its life. Analysis of the processes' phases presented on Fig. 1.2 highlights:

(1) Clark and Fujimoto's [7] proposal is focused on execution (engineering), and only partially (in gray) considers the interface with manufacturing and ramp-up;
(2) Wheelright and Clark [8], though keeping the execution focus, consider a higher participation on the ramp-up;
(3) Ulrich and Eppinger [3], on the other hand, explicitly consider the planning (and not implicitly on the conceptual phase);
(4) Anderson [9] includes the product follow-up after the market introduction;
(5) Cooper [10] describes in detail the financial and market concerns;
(6) Rozenfeld et al. [2] broaden the PDP scope to encompass the whole product life-cycle, including the developments that will evolve and keep the product competitive in the market until its discontinuity; and

(7) The value creation framework proposed by Murman et al. [1], though not a development process, resembles the PDP models very much, making a link to the lean philosophy.

(8) This book's PDP, which is defined in sequence and further described in Chap. 9.

Product Development Process (PDP): The set of activities beginning with the perception of a market opportunity aligned to the company's competitive strategy and technical capacity, and ending in the production, sale, and delivery of a product, while considering all aspects that will evolve and keep the product competitive in the market until its discontinuity.

Product: All the results from the PDP, not limited to physical products, but also encompassing services, product-as-service, and even complete value chains, which are aimed to fulfil the customer and user needs.

By considering the results from the PDP as "product", whatever is the shape they take, the PDP becomes more aligned to the lean philosophy. As presented in Chap. 4 and further detailed in Chap. 10, a Lean Product Development Process aims to fulfill the value pulled by the stakeholders. Depending of the chosen value delivery architecture (see Chap. 11), this value is delivered through physical products, services, or any mix of product and services. Sometimes the defining of a completely new value chain and/or business model is necessary to deliver the pulled value.

By considering the complete product lifecycle the PDP takes into account all aspects the product is going to face through the lifecycle stages. This approach is aligned to reducing the total cost of ownership and increasing the total revenue of servitization, in the case of product-service systems, where a product-service system is a marketable set of products and services capable of jointly fulfilling a user's needs.

This view of the PDP also embeds a product management mindset, where the further evolution of the product after the sale or market launch is part of the PDP. This is also aligned with the lean philosophy, once the value pulled by the stakeholders might change through the time (due to market changes, technology evolution, etc.) and the offered "product" should evolve accordingly.

The icon of a funnel (Fig. 1.3) has also been used as a visual depiction of the PDP. It works well because it implies that product development is, in fact, a refinement process that takes us from the earliest stages of a project—with a lot of fuzzy ideas and fuzzy thinking—to the final stage of new product launch.

The funnel metaphor is also very aligned to the lean PDP, which uses the Set-based Concurrent Engineering (SBCE) to maintain product design options through the PDP, instead of choosing a particular option to pursue from the beginning. Although this option is carefully chosen from the other possible alternatives through cost-benefit and risk analysis, the point-based approach often implies in

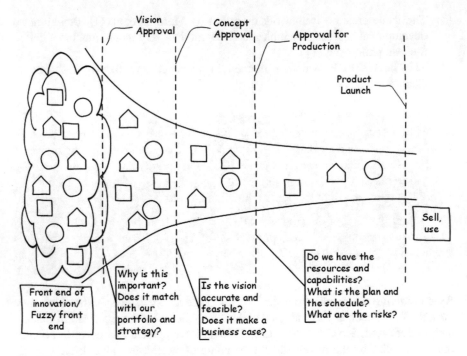

Fig. 1.3 Product development funnel

rework cycles, which might disrupt the whole development portfolio (see Chaps. 9 and 11 for more details about SBCE).

1.2 The Product Development Process Particularities

The PDP itself is a creative, innovative, interdisciplinary, dynamic, highly coupled, massively parallel, iterative, communication-based, uncertain, and risky process of intensive planning and activity [11]. Consequently, a wide spectrum of variables can affect its success, and, not surprisingly, over time, over budget and low quality are commonplaces on PD projects.

Defining or improving a PDP should be proceeded by a reflection on how these particularities affect your own company. Different markets, business models, culture, etc. might lead to distinct impact from these particularities. This is also true with different development centers in the same company, since the organizational culture from each center might be different (i.e. globally distributed development). As a consequence, these variations should be taken into account when defining a company-wide and global PDP.

1.2.1 Uncertainty

Uncertainty is the knowledge gap between the supposed and the verified characteristics, and lasts while the development is in progress. The uncertainty is directly related to risk [12]:

- **Performance risk**: Uncertainty in the ability of a design to meet desired quality criteria (along any one or more dimensions of merit, including price and timing), and the consequences thereof.
- **Schedule risk**: Uncertainty in the ability of a project to develop an acceptable design (i.e., to sufficiently reduce performance risk) within a span of time, and the consequences thereof.
- **Development cost risk**: Uncertainty in the ability of a project to develop an acceptable design (i.e., to sufficiently reduce performance risk) within a given budget, and the consequences thereof.
- **Resources/Technology risk**: Uncertainty in capability of the resources (including people) and technology to provide performance benefits (within cost and/or schedule expectations), and the consequences thereof.
- **Market risk**: Uncertainty in the anticipated utility or value to the market of the chosen "design to" specifications (including price and timing), and the consequences thereof.
- **Business risk**: Uncertainty in political, economic, labor, societal, or other factors in the business environment and the consequences thereof.

This gap might lead the whole development into wrong assumptions, causing frequent estimate failures and rework cycles. The earlier in the product development process, the higher the uncertainty, thus making important decisions is based on assumptions.

1.2.2 People-Based

PD is a people-based activity, where each person has his/her own culture, values, personality, etc., and may present unpredictable behaviors, i.e., a "box of surprises."

As a consequence, the time it takes to perform an activity will not likely be the same whether it is done by different people or if the same person does the same activity on different occasions. Product development processes will always embody statistical fluctuation during their execution. Higher deviations from the average execution time are expected when dealing with new processes, innovative products, and unmastered technologies.

1.2.3 Ambiguity

Ambiguity means the existence of multiple conflicting interpretations of the information held and required which leads to a lack of consistent information. While uncertainty leads to the acquisition of objective information and answer specific questions, ambiguity leads to the search for the meaning of things. The customer needs or project goals might not be clear, and the information that flows during the development often carries a level of ambiguity and uncertainty.

1.2.4 Non-linearity

Product development is not a sequential and linear activity. The more innovative the product, the more complex it is to find a suitable architecture in the solution space. Therefore, the PDP is an iterative process comprised of:

- **Iteration**: Iteration is the procedure by which repetition of a sequence of operations yields results successively closer to a desired result. Iteration can be planned (iterative process) and unplanned (rework). Too complex/poor interface design may lead to more iteration. The higher the number of unplanned iteration cycles the worse the overtime becomes.
- **Interruption**: Critical design issues, trivial questions, unplanned communication, multitasking, etc. always arise during the development. Though natural, the higher the interruption level on the development projects the worse.
- **Changes**: Nothing ever happens exactly the way it was planned (changing requirements, resources unavailability, etc.). High change rates compromise the development progress.

1.2.5 Complexity

The PDP also has to face complexity at multiple levels: the product itself, the development process, and the performing organization (development teams included) [13]:

- **Product complexity**: Customers request products that are more and more complex themselves. The product development scope includes not only the final product itself, but also its life-cycle processes and the performing organizations of these processes.
- **Processes/tools complexity**: The increasing number of processes and tools and the challenge to keep them integrated at some level creates issues for effective and unambiguous communication.
- **Structure complexity**: The performing organization's structure is becoming more and more complex to be able to deal with increasing product and process complexity, as well as to adapt to global markets and distributed development. The bigger, more distributed, and more multidisciplinary the development team is, the more intensive the need is for communication and coordination to keep the work aligned.

As the complexity of the product increases, the number of different expertise needed to design it also increases. A cooperative environment with mutual help and knowledge sharing is paramount to the development success. This poses great management and product integration challenges.

These particularities help us understanding why consistently succeeding in product development projects is challenging. Any high performance product development process should tackle these aspects in an integrated way. The process we describe in Chaps. 9–13 act in this way.

1.3   Product Development Performance Drivers

Product development is indeed a complex endeavor. The PDS can be understood as a network with multiple dimensional and highly interconnected processes where feedback-loops cross these multiple hierarchical levels. As a result, there are several drivers that impact the performance of development projects [14]. We divided these drivers into four groups (Fig. 1.4) which are detailed as follows. A complete description of each group's performance drivers categories and subcategories is presented in Appendix A.

The importance of understanding these drivers is to identify their presence in any particular development project and/or Development Organization. They explain the current product development performance, and are a good start to any process improvement effort, as we are going to present in Chaps. 5 and 6.

1.3.1   External Environment

The external environment group includes all the issues that originated outside the PDS and the parent organization. Though the company has little or no power to influence the environment, some particularities of this environment can directly

Fig. 1.4   Groups of drivers

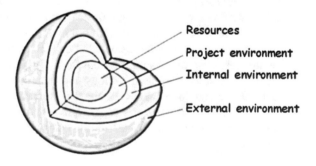

Resources

Project environment

Internal environment

External environment

affect the shape of the enterprise's PDS and its success. The external environment is divided into two categories:

1. **Market**: Even though the company can perform research and prepare its products for the market, the market itself is outside the company's boundaries, and consumer decision, globalization, and product lifecycles are some aspects that might influence the product success.
2. **Business**: The category of business includes all the external factors except the market itself. Instabilities on the business include change on the political, economic, and labor scenarios.

The external environment influence in the PD explains the great uncertainty that any PD project faces. The longer the development project takes, the bigger are the chances that the market or the business might change in a way that impact the development; therefore causing rework cycles or even turning the complete project obsolete.

1.3.2 Internal Environment

The internal environment includes everything that is outside of the PDS but is still within the boundaries of the parent organization. In most of the companies the PD department (if any) or the PD team are part of a greater organization. As a consequence the PD structure is influenced by this larger body. Dealing with the internal environment requires from the PD team leader good knowledge of the organization culture and policies, and good communication and negotiation skills.

The internal environment is divided into five categories of aspects that can have an impact on the PDP performance by not giving the necessary support to its management and execution:

1. **Organizational culture**: The company's values, beliefs, assumptions, perceptions, behavioral norms, artifacts, and patterns of behavior create the organizational culture. Therefore, it plays a critical role in how the PDS is really structured and executed, sometimes in ways different than the company's standards.
2. **Corporate strategy**: Objectives, purposes or goals, main policies and plans for achieving those goals, the range of business the company is to pursue, the kind of economic and human organization it is or intends to be, and the nature of the economic and non-economic contribution it intends to make to its shareholders, employees, customers, and communities. Unclear strategies or the misalignment between the corporate strategy and the development needs and goals is a factor that can reduce the development performance.
3. **Organizational structure**: Responsibilities, authorities, and relations organized in order to enable the performing of organization functions, including the product development.

4. **Business functions**: This category considers the issues between the product development and the other business functions in the company such as human resources, sales and marketing, research and development, production/operations, customer service, finance and accounts, and administration and information technology.
5. **Supporting processes**: These are processes that permeate several business functions, such as process improvement, training and knowledge management.

1.3.3 Project Environment

Project environment encompasses all the product development management and execution activities and is divided into six categories: initiation, development planning, execution management, development control, communication, and development execution. Issues on these aspects will directly impact the PDP performance.

1. **Initiation**: Defines and authorizes the development; guarantees the alignment between the development and the corporate strategy through clear and feasible objectives.
2. **Planning**: Defines and refines objectives and plans the course of action required to attain the objectives and scope that the project was undertaken to address.
3. **Execution management**: Integrates people and other resources to carry out the planned project for the project.
4. **Development control**: Regularly measures and monitors progress to identify variances from the project management plan so that corrective actions can be taken when necessary to meet project objectives.
5. **Communication**: Includes all the issues that could interfere with an effective exchange of information.
6. **Development execution**: Includes all the issues of effective engineering, its subcategories are: requirements development, technical solution and integration, and verification and validation.

1.3.4 Resources

This group considers the issues related to people, tools, and standards involved during development.

1. **People**: People execute the development itself; they must have the proper knowledge, experience, and skills to positively contribute to the product development success.
2. **Tools**: Tools are used by the people to perform their development tasks; they not only must be adequate to each task individually, but they also must be at

some level integrated between themselves, allowing a smooth development flow.

3. **Standards**: Standards guide the work. Good standards, on the one hand, help reduce the variability of the development process, increasing the quality of each task outcome and the development success as a whole. Bad standards, on the other hand, provide misguidance and confusion by either requesting the wrong deliverables (do the wrong work), or by suggesting a non-coherent or badly defined set of processes (do the work incorrectly).

1.4 Product Development Metrics

Once the drivers to product development low performance are understood, it is important to define how to measure this system and determine how the environment influences the results of this measurement. There are seven categories of indicators to the Product Development System:

Product quality: Product quality has several interpretations ranging from design quality; enterprise capacity to produce the product according to the design; conformance (reliability in use); delivery of the scope; fulfillment of the company's strategy (not only bounded by the initial project scope); and simply the satisfaction of all stakeholders' needs, or rather, delivering all the expected value.

Product business case: One important aspect about product quality is that the quality needs perceived at the beginning of the development and the actual needs when the final product is delivered might differ. The customers, the market, the laws, etc. might change and impact the product acceptance. Therefore, keeping track of how strong your business case is through the development project is paramount.

Development time: The development project must deliver the product scope on time. Development (lead) time measures how quickly the company can move from concept to market, and the enterprise responsiveness to the competitive forces and the technological evolution. Short development lead times increase the frequency of new products introduction.

Product cost and Development cost: The development project must also deliver the product scope within the budget. Both product cost and development cost are of importance; the former constrains the enterprise profit according to the volume and selling price, the later constrains the return on investment and the enterprise capacity to do several developments at the same time. The product cost includes material, labor, and the needed production tooling, as well the incremental costs to produce additional units. The development cost includes all the development expenditures.

Development productivity: The aspects related to the product to be developed, as well as to the development must be followed in order to guarantee that "what" we are developing, its cost, and its delivery date, will always sustain a viable business case. A product that does not fulfill the market needs, at the right cost, and at the correct market window should not have been developed.

Productivity determines the level of resources required to take the project from concept to commercial product. This includes hours worked (engineering hours), materials used for prototype construction, and any equipment and services the company may use. Productivity has a direct though relatively small effect on unit production cost, but it also affects the number of projects a firm can complete for a given level of resources.

Development capability: The accumulated knowledge/experience from previous projects that increase the productivity of future projects is included in development capability.

Some of these categories are related to product indicators, while others are related to process indicators. The product indicators measure if the right product is being developed; the process indicators help understanding it the product is being developed in the most effective way. Product quality, product business case, and product cost are product indicators categories. Development time, development cot, development productivity, and development capability are process indicator categories.

This division into product and process indicators influence how the continuous improvement in the PD context (see Chap. 6). Indeed the PDP is a continuous improvement process itself, once it gradually improves the developed product indicators. A low performance PDS is the consequence of issues that negatively impact the performance indicators of product quality, product business case, product cost, development time, development cost, development productivity, and production capability.

Several metrics can be used to support these indicators. Appendix B presents some commonly used Product Development Program metrics. Application of the SMART criteria is one widely used way to choose the metrics that fit your company:

Specific: Ensure that program metrics are specific and targeted to the area being measured.

Measurable: Make certain that collected data is accurate and complete.

Actionable: Make sure the program's metrics are easy to understand and clearly chart performance over time so that decision makers know which direction is "good" and which direction is "bad."

Relevant: Include only what is important and avoid metrics that are not.

Timely: Ensure that program metrics produce data when it is needed.

The list of commonly used program metrics presented in Appendix B shows that there is a myriad of possible metrics to choose. When choosing which metrics to use, the company should look into the ones that make sense to its particular needs and which of them they are capable to measure. A common mistake is choosing a great number of metrics (some of them even redundant) and facing the wasteful effort of measuring all of them.

Finally metrics are incremental; each level of the organization aggregates its metrics to the upper level, thus turning feasible the management. Take a complex development program where the product has several subsystems, an aircraft development for instance. The group in charge of each subsystem has its own set of metrics, and the program manager has aggregate metrics that help him managing the complete development. Only when something goes wrong with a particular subsystem, that the program manager goes deeper into its individual metrics.

 1.5 A Practical View

As in any complex system, the PDS cannot be described by analyzing its parts separately; the final system behavior emerges from the interaction among its parts. In practice, the analysis of any working PDS must be done by checking the interfaces between the system and its environment, and among its constituent parts.

The most important (and easier) issues to be perceived are related to the PDS outputs. Costumer complaints, the need of recalls, losing market share, etc. are symptoms of a low performance PDS. It is important to "ask why" and go deep into understanding the perception of the problem (Fig. 1.5). We are not trying to find the root causes yet, but addressing and trying to define the real problem. Going deeper into the PDS and finding the issues among the system's parts is the path to find the wastes (see Chap. 5) and the root causes.

If any issues in the PDS inputs are found, not only must its causes be understood, but also whether the PDS can be improved in order to have more robust capability to handle input variance.

A way to apply the contents of this chapter in practice is:

1. Look at your actual PDP, how does it compare to the processes presented in Fig. 1.2? Does it encompass the whole product lifecycle? Lean Product Development Processes take into account the whole lifecycle, if yours does not it is an opportunity to expand it by including integrated product design and development strategies and techniques as presented in the next chapter.
2. How do you consider the "product"? Is it the consequence of delivering the pulled value or some predefined result that you push to the market/client? It is almost impossible to have a certain solution idea when starting a new product development project. This is not bad in essence, since it gives focus about the benefits it will produce. You should detach from this particular solution though and concentrate in the value it is going to deliver. The next step is considering what other product-service architectures could the lever the same (and even more) value.
3. Read again the PD particularities and try to identify them in the PD projects you recently executed. Can you see some of them? Are they understood by the development team? This exercise has the potential of helping you identifying the particularities of the PDP in the market you are inserted, and improving your process.

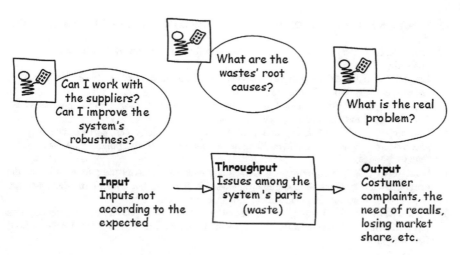

Fig. 1.5 Finding issues in the PDS

4. Check the PD performance drivers (also check Appendix A). Try identifying which of the categories and subcategories are present in your company. By identifying them and understanding their root causes, you can make real improvement in your process. This work is closely related to what is presented in Chaps. 5 and 6.

Indeed, it is a great challenge to design and develop winner products. As a consequence, the PDP has constantly evolved through time in order to address the low performance drivers.

The next chapters show this evolution from serial PD, to integrated PD and, finally, to lean PD.

References

1. Murman et al (2002) Lean enterprise value: insights from MIT's lean aerospace initiative. New York, NY, Polgrave
2. Rozenfeld H et al (2006) Gestão de Desenvolvimento de Produtos. São Paulo, Editora Saraiva
3. Ulrich K, Eppinger S (2004) Product design and development, 3rd edn. McGraw-Hill, New York
4. Bertalanffy L (1968) General system theory: foundations, development, applications. George Braziller, New Yor
5. Blanchard BS, Fabrycky WJ (1988) Systems engineering and analysis, 3rd edn. New Jersey, Prentice Hall
6. Kotler P (1998) Administração de marketing, 5th edn. São Paulo, Editora Atlas
7. Clark K, Fujimoto T (1991) Product development performance: strategy, organization, and management in the world auto industry. Harvard Business School Press, Boston

8. Wheelright S, Clark K (1992) Revolutionizing product development. The Free Press, New York
9. Anderson DM (1997) Agile product development for mass customization. Irwin Professional Publishing, Chicago
10. Cooper RG (2001) Winning at new products: accelerating the process from idea to launch, 3rd edn. Perseus Books Group, Cambridge
11. Negele H et al (1999) Modelling of integrated product development processes. In: Proceedings of the 9th annual symposium of incose. [S.l.]: [s.n.], 1999
12. Browning T (1998) Modeling and analyzing cost, schedule, and performance in complex system product development. Technology, management and policy. PhD thesis, MIT, Cambridge, MA
13. Loureiro G (1999) A systems engineering and concurrent engineering framework for the integrated development of complex products. PhD thesis, Department of Manufacturing Engineering, Loughborough University, Loughborough, UK
14. Pessôa MVP (2008) Weaving the waste net: a model to the product development system low performance drivers and its causes. Lean Aerospace Initiative Report WP08-01. MIT, Cambridge, MA

Chapter 2
Integrated Product Design and Development

This chapter discusses the evolution of product development approaches through time. It reviews the serial or sequential approach that was adopted by the companies under the influence of Industrial Revolution and Fordism. Then the Integrated Product Development (IPD) approach is presented and discussed. It's worth mentioning that IPD was influenced by the Computer Integrated Manufacturing (CIM) proposal that emerged in the 80s as an evolution of the Ford manufacturing system. IPD keeps the benefits from the former approach (shorten price, shorten time-to-market, augmented quality) while fixes its shortcoming such as reworks, lack of communication amongst technical areas etc. IPD prescribes the structuring of two main pillars, namely, multifunctional or IPD teams and DFX (Design for eXcellence) design tools. After presenting some practical examples of the usage of DFX design tools, this chapter introduces the novel concept of integrative design variables (IDV): there is a target value associate to them; they are affected and affect most of the design decisions and their meaning is easy to grasp. Cost, weight, center of gravity are IDV examples. The IPD concept goes far beyond standard products such as cars, aircrafts and washing machines. At the end of the chapter you'll find the IPD applied to academic or technical assessment.

2.1 Introduction

As discussed in the previous chapter, Product Design and Development can be seen as a process and, consequently can be modeled. Figure 2.1 depicts the simplest possible model for such a process.

Although simple, the model contains the main activities of the product development process (PDP), namely, needs identification, synthesis of the product, and evaluation of the design alternatives for the product.

The PDP ought to start with the customer needs identification. Sometimes these needs are presented in a broad way, such as the need for reducing atmosphere CO_2 emission from the airplane, and other times in a very strict sense, such as defining an automatic procedure for installing aircraft rivets.

© Springer International Publishing AG 2017
M.V.P. Pessôa and L.G. Trabasso, *The Lean Product Design and Development Journey*, DOI 10.1007/978-3-319-46792-4_2

Fig. 2.1 The simplest model
of product development

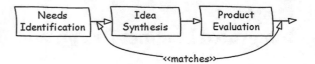

Some people think that PDP is all about synthesizing ideas and conceiving new products. Synthesis plays an important role within the product design and development, but is merely one activity of the PDP and has equal weight as the others.

The evaluation activity verifies whether the synthesized product meets the needs identified at the very beginning of the process. If not, a design loop is established up to the point that the needs are met. Evaluation of the product alternatives is as important as the needs identification and the product synthesis. How engineers have approached these activities through time has defined Product Design and Development evolution.

2.2 Sequential Product Design and Integrated Product Design

During the Industrial Revolution of the 18th century, a company that designed and built steam engines had argued that its engine was better than its counterpart. Naturally, the other company had the opposite opinion. Quarrels like this only ceased after the publication of the thermodynamics laws that were used as quantitative criteria for evaluating the best design alternative for the steam engine.

How do we evaluate design concepts? The first move for many of people is to base it upon the product functionality. Suppose you are given the following requirements (needs): design a product which is capable of lifting a 500 kg block of steel to a height of 1 m from the floor, moving it along a 3 m straight path at a speed of 0.5 m/s and lowering it back onto the floor. What would come to your mind? Steel cables, hooks, electric motors, brakes, axles, pulleys etc. Putting all the components together, your product would work, or should we say it would *function* because it resembles a mechanical hoist.

In this example, have you thought about the best way of assembling the components? How could it be easier for the maintenance personnel to execute their jobs? How should the components be manufactured in order to facilitate access to the tooling? These are questions beyond the functional evaluation which need answered during the product life cycle.

In the serial product development (SPD) (Fig. 2.2a), on the one hand, only the functionality of the product is taken into account during synthesis. Regarding the previous example, if the product designed does lift the 500 kg steel block, moves it, and puts it back on the floor, it does function! Manufacturing, assembly, maintenance issues and so forth are solved later by somebody else.

On the other hand, in the integrated product development (IPD) approach (Fig. 2.2b), the requirements from the product lifecycle areas such as: design,

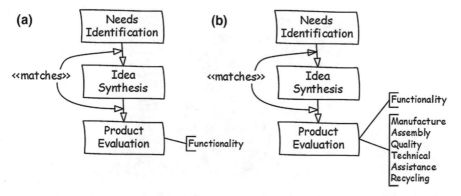

Fig. 2.2 Serial (**a**) and integrated (**b**) model of PDP

manufacturing, assembly, maintenance, disposal, and so on, are also considered, weighed, discussed, and balanced at the conceptual phase of the product development. [1] As a result, the outcome from integrated product development is a product which is designed not only to work, but also to be easily and cheaply manufactured, assembled, tested, maintained, and recycled.

By comparing the models of the serial product development and the integrated product development, one can realize that the latter encompasses the former. IPD expands the horizon of the product evaluation by trying to take into account all the technical areas and phases the product goes through during its lifecycle.

2.3 Serial Product Development

Before the Industrial Revolution, there existed the most–ever–integrated product development. Think of an artisanal shoemaker. He knew how to design the shoes and he mastered all necessary tools and tooling for manufacturing the shoes which would fulfill all the functions and needs from his neighborhood. The shoemaker knew the tastes of his customers and would ask from time to time whether a small repair was due. Market needs, product conceptual design, manufacturing, assembly, and maintenance were integrated in the shoemaker's head at the speed of synapses (Fig. 2.3).

With the Industrial Revolution also came the division of work into specific technical areas. The product development process mirrored the serial production line, thus adopting the serial approach as well.

One might rightly argue the benefits brought by the industrial revolution: product costs reduction, production increase and a raise in quality standards. These benefits still exist today, but it stands to reason that the mass production era split the technical areas of the product development process into separate departments (i.e., silos), based upon highly skilled people within them, but with almost no interaction among them.

Fig. 2.3 Synergy of the integrated product development

Within the PDP, a typical manner of work by the multiple departments is to finish their jobs as quickly as possible and throw them over the "wall" to the next department. Suppose that the designer from Fig. 2.4 "threw" a blueprint in which a 3.3 mm diameter hole has been drawn. As soon as the blueprint lands in the manufacturing department, the technician will realize that there are no 3.3 mm commercially available drills. Naturally, the manufacturing technician could not choose an available drill whose diameter was close to that specified by the design department. Then the project oscillation begins: the manufacturing department writes down a design change request that will be analyzed and eventually implemented after some interaction loops.

The barrier metaphorically represented in Fig. 2.4 extends to all the areas participating in the PDP, creating a great challenge to integration.

Rescuing the integration of the PDP is therefore a challenge, but this should not hamper the attempt to achieve it. One can find the motivation to pursue it by looking at the negative consequences of the serial product development:

- Production is not considered in the conceptual phase of PDP but only at the very phase of production where the product modifications, if needed, are more difficult to implement and costly.
- Product data is fragmented so that each technical area has its own product data representation.
- Product development is driven by milestone dates associated to each development phase; thus, putting pressure on technical specifications and drawings release. As a consequence, few design alternatives are evaluated.

Fig. 2.4 Organizational barriers for the integrated product development approach

Fig. 2.5 Typical behavior of the product development process

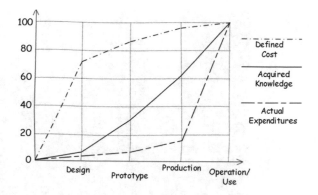

These negatives consequences could be used for justifying the replacement of the serial to the integrated approach for product development. However, a stronger and more eloquent reason could be drawn from Fig. 2.5.

The PD life cycle phases are represented at the abscissa axis. The ordinate axis shows the percentage magnitudes of three important variables within PDP, namely, defined cost to implement a given PDP phase, knowledge acquired about the product, and actual incurred cost of the product defined.

Figure 2.5 draws attention to the 75 % mark of the defined cost regarding the conceptual (design) phase of PDP. This means that 75 % of the overall forthcoming cost of the product is defined at the conceptual phase of PDP. It is not difficult to figure out the causes: the designer has to define the shape, geometry, and features of a product which are strictly related to the manufacturing process. In addition, the designer ought to define, but not yet buy, the materials of the components and parts. To complete the product specification, the engineer/designer has to define geometric and dimensional tolerances of the components and parts as well as define the surface finish.

The gap between knowledge and cost decisions implies that many decisions are made based on wishful thinking, therefore causing rework and correction loopbacks during the remainder of the PDP and product life cycle.

2.4 Integrated Product Development: A Rescue Movement

Integrated product development is all about rescuing the interaction among the technical areas concerning the product in which requirements are taken into account and balanced for the benefit of the product. This is a "rescuing" movement because this integration once existed and was lost. However it is not possible to rescue the integration as it was at the artisanal production level. It is no longer possible for a single person to keep with all the information and have all the knowledge needed to consider all aspects of the lifecycle of a typical complex product of current times, as shown in Fig. 2.6. Even if we drop the complexity of

Fig. 2.6 A highly complex product: EMBRAER KC-390. *Source* Disclosure Embraer

Fig. 2.7 Knowledge
production versus knowledge
acquisition

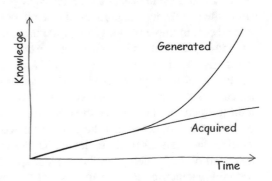

the product to a cell phone, the challenge for recreating the integration environment remains.

Indeed, knowledge generation poses a great challenge for successfully rescuing the integration of the product development process. At this very moment, a great amount of brand-new knowledge is being produced all over the world so that it is literally impossible to catch up to such a pace as Fig. 2.7 represents.

Considering what was presented in Fig. 2.5, by comparing the knowledge and expenditures curves within the conceptual phase of PDP, one can realize that most of the budget commitment is taken with a low degree of knowledge. These decisions should take into account not only aspects from the conceptual phase, but from other phases, such as manufacturing and assembly, as well. Thus, the decisions and definitions of the conceptual phase ought to be taken with a higher degree of knowledge as depicted as in Fig. 2.8.

One of the objectives of IPD is to increase the knowledge of the product at the earliest phase of the PDP, and supporting the decisions that must be taken at this moment. The actual expenditures line indicates that investments should be made in

Fig. 2.8 A new proposal for PDP: knowledge build up through integration

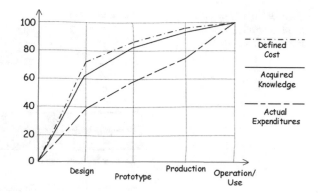

order to create this knowledge. To gather information and requirements of dimensional and geometric tolerances, for instance, the company might contract and pay for consultancy in that field. It's worth mentioning that the percentage of 75 on the defined cost remains the same in the IPD scenario because the decisions about product geometry, materials, tolerances, and surface finishing need to be taken regardless of the increase in knowledge about the product.

If we analyze the expected time interval from the early phases on PDP (note that Figs. 2.5 and 2.8 do not include time), considering both the SPD and the IPD, the time interval needed to accomplish the conceptual phase in the latter is greater than in the former (Fig. 2.9). This is the consequence of the early exchange of information among the different areas which aim to reduce the total design and development lead time.

While major design changes were anticipated and solved at the conceptual phase where the changes are easier and cheaper to implement, some design changes might still occur at the IPD remaining phases. However, these changes are significantly less important than those discussed and solved at the conceptual phase.

These duration's expectations, as shown in Fig. 2.9, pose a managerial dilemma of IPD. How could a company be certain about the expected decrease of the product development lead time?

Take, for instance, the aeronautical sector. Even though there is public data about expected lead time reductions from 48 to 36 months during aircraft development, the manager will have to wait 3 years before becoming certain of the IPD investment return. In the meantime, the manager will receive all the pressure to present tangible results, while keeping a *"Festina lente"* (make haste slowly) attitude. Overcoming this managerial dilemma is one of the challenges for a company which is used to the serial approach to change into integrated product development.

A way to surmount the managerial dilemma is to select a pilot development of a product that is simpler and quicker to implement than the company's main product. For instance, an aircraft manufacturer could choose a fuel tank to initiate the

Fig. 2.9 The integrated product development managerial dilemma

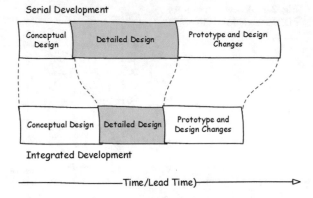

IPD approach. After gaining confidence during this pilot project, the experience and knowledge could be extrapolated to other systems and then to the company's main product.

There are, essentially, two main resources required to implement the IPD approach, namely, multifunctional design teams and IPD tools.

2.5 Integrated Product Development Teams

Suppose you look at a bus stop and see a gathering of people. I what you see a group or a team? Certainly, it is a group of people because one person might go to the town center, another to suburb, and so forth. Then, some people get on the bus heading to the town center. Are the people inside the bus a group or a team? Certainly, it is a group as one person might stop by the library, another to train station, and so forth. Finally, you see some people from that bus coming out of it to try to fix an engine breakdown. Are the people trying to fix the engine a group or a team? Certainly, it is a team. From this simple example one might figure out that the four characteristics of a team are: mission, commitment, complementary capacity, and ephemerality as shown in Fig. 2.10.

The mission of the IPD team is to assure that the requirements of all product development phases are evenly represented in the IPD's conceptual design phase. All people from the IPD design team should be committed to obtaining the best possible balanced results for the product, even if that means giving away some of his/her technical area expectations [2]. Complementary capacity is achieved by having representatives from all PD technical areas on the IPD team, such as: marketing, design, manufacturing, assembly, maintenance, etc. An IPD design team is ephemeral because after finishing a given product development, that particular IPD team ceases to exist.

Ideally, all technical areas from the product lifecycle phases are represented in a typical design team meeting and a number of engineering tradeoffs are raised, discussed, and solved. For instance, the choice of a certain electric spindle for

Fig. 2.10 Multifunctional design teams

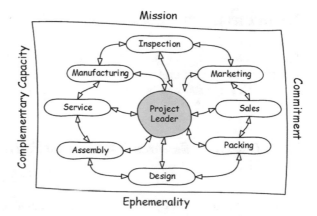

a robot end effector might suit the power requirements for the drilling operations. On the other hand, that very spindle could jeopardize the weight payload of the robot. In another example, a person from manufacturing would argue that the product geometry would be better that way in order to avoid reorientation of the part. A person from marketing would argue that the geometry just proposed by manufacturing would not sell a piece. The adequate spindle should suit both requirements. At the end of the meeting, nobody leaves either "100 % happy" or "100 % unhappy."

It is the role of the project leader to ensure the team's focus on the mission and achieve a balanced result. The specialists, while they have deep vertical knowledge in their subjects, must have the maturity to explore the horizontal knowledge which is how each subject can interface in order to leverage the others and benefit the product as a whole (Fig. 2.11).

Fig. 2.11 Vertical and horizontal knowledge in IPD design teams

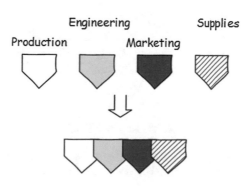

2.6 Integrated Product Development Tools

This section presents a quick overview of some IPD tools. We show the benefits brought from integration to the product development process without going into much detail about the step-by-step use of the tools.

2.6.1 Design by Features (DbF)—as a Potential IPD Design Tool

Design by Features (DbF) is a CAD resource; although it is not an IPD tool, it gives a good example of possible tool adaptation to support the cooperation among the design and the manufacturing teams.

DbF was developed in the early 1990s to replace the cumbersome way of drawing manufacturing features such as holes, pockets, edge fillet, and so forth in a CAD design. Prior to DbF, the CAD designer had to make a Constructive Solid Geometry (CSG) approach, drawing two solids, a block and a cylinder, aligning them, and making a Boolean subtraction to draw a hole (Fig. 2.12a). Alternatively, he/she could use the Boundary Representation (BRep) approach by drawing the whole surfaces and assembling them afterwards (Fig. 2.12b).

For drawing the same hole using the DbF approach, a CAD designer chooses the feature <hole> from a drawing pallet, defines the hole type, for instance, <blind> as well as the hole dimensions: <diameter> and <depth> as shown in Fig. 2.13 and indicates the place the hole should be on the workpiece.

This is a pure CAD task assisted by DbF. Suppose that the engineer or designer chooses the hole's diameter as 10.37 mm. The CAD will draw the holes all the same. When the CAD file is sent to manufacturing shop floor, the drill operator will not find a commercially available 10.37 mm drill for drilling the hole as specified by engineering. The drill operator does not have the authority to decide upon a different hole's diameter based upon the commercially available drills. Therefore, a communication protocol has to be established between design and

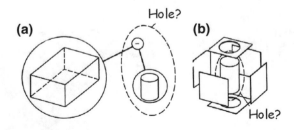

CSG: Constructive Solid Geometry   BRep: Boundary Representation

Fig. 2.12 CAD approaches for drawing manufacturing features. **a** *CSG* Constructive solid geometry, **b** *BRep* boundary representation

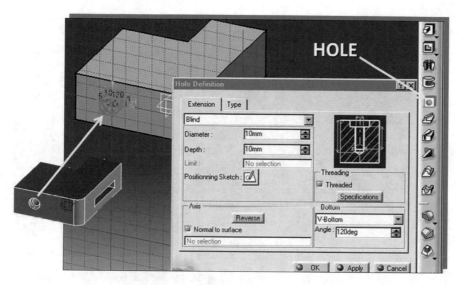

Fig. 2.13 Drawing a hole using the DbF approach

manufacturing areas in order to decide the final diameter of the hole. This value could even be 10.37 mm, but it will require a special drill that costs more than the commercial counterpart. Nevertheless, this is an example of oscillation in the product development process characterized by a loop that consumes valuable time without adding value to the product.

Suppose the engineer or designer is presented a set of commercially available drills, as soon as he or she selects the <diameter> scroll bar in the CAD screen as shown in Fig. 2.14.

The IPD-DbF design tool has the <diameter> scroll bar locked to a drill table so that the designer is required to choose one of the commercial available drills. This is a true design-manufacturing integration accomplished in a very efficient and clever way as the designer does not leave his/her work environment to search for information, and the drill operator will do his/her job as soon as he/she is required to. At the very end of that table, after all commercial available drill choices, a blank field can be shown to deal with special cases such as the 10.37 mm drill.

2.6.2 Knowledge Based Engineering (KBE)—a Truly IPD Design Tool

KBE is a computer-based design environment where the design intent can be captured, executed, and disseminated through a company. Suppose an engineer is given the task of dimensioning spars and ribs of an aircraft wing as shown in

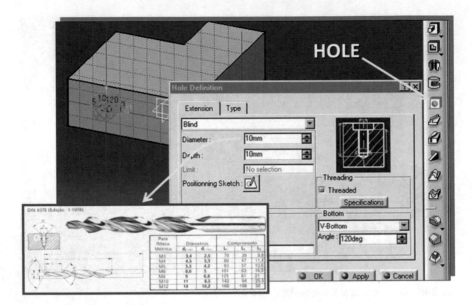

Fig. 2.14 DbF adapted as an IPD design tool

Fig. 2.15 Spars and ribs dimensioning of an aircraft wing

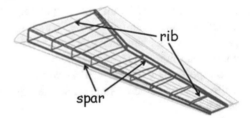

Fig. 2.15. Both the leading edge spar and the trailing edge spar are…spars! Nevertheless the engineer has to repeat the dimensioning procedure for both spars taking into account the differences of geometry, load conditions, and assembly docking features. The same situation happens for the 10 ribs shown in the picture.

To overcome the burden of repeating the dimensioning procedure as exemplified, KBE turns the definition of a specific product design process into creating rules, activities, and decisions that a skilled engineer would follow to accomplish the product dimensioning and design. Therefore, the written design procedure would be created by a person (the same engineer who wrote it or somebody else) and all the spars and ribs would be dimensioned, designed, and drawn. The creation of the written procedure looks like a CAD parametric window (Fig. 2.16), where the user inputs some information about the part he/she wants to design, such as part location, space among the parts, maximum load upon a part etc.

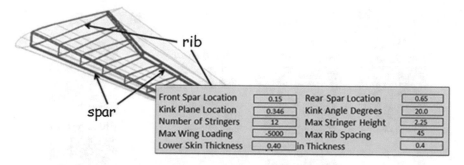

Front Spar Location	0.15	Rear Spar Location	0.65
Kink Plane Location	0.346	Kink Angle Degrees	20.0
Number of Stringers	12	Max Stringer Height	2.25
Max Wing Loading	-5000	Max Rib Spacing	45
Lower Skin Thickness	0.40	in Thickness	0.4

Fig. 2.16 Spars and ribs dimensioned automatically by the KBE engine

The KBE design environment is like a blank sheet of paper where a specialist applies his expertise of how to design and dimension a given part. The key part of KBE is named *generative model*, where the design intent, dimensioning procedures, rules, tables, and other contents can be linked to finite element analysis, cost analysis, manufacturing and assembly restrictions, guidelines, and evaluations. Due to this additional content, KBE has the potential to establish integration among the technical areas of the product life cycle. The dialogue between design and manufacturing, for instance, might be accomplished through a written procedure of the generative model of KBE. This is exemplified by a practical industrial case depicted in Fig. 2.17.

In the serial tube design and manufacturing approach (Fig. 2.17a), the designer draws a 3D tube to meet functional requirements such as to connect two ends of the air conditioning system. To accomplish that, a series of bends and curves need to be drawn and modeled. After finishing the design and modeling of the tube, a manufacturing engineer checks whether the bending machine is capable of bending the tube with the angles specified by the designer. If all the angles are feasible to be manufactured, the part number is approved; otherwise, the 3D drawing of the tube returns to the designer who, by his/her turn, corrects the angles. Once the part number is approved, a set-up operator inputs the necessary data to run the bending machine. Then the bend machine operator finishes the process and the tube is ready to be installed. It's clear that this design process has several opportunities for improvement, mostly related to the elimination of the design oscillation phenomenon already described herein.

An integrated tube development approach based upon the KBE engine is pictured in Fig. 2.17b.

In this approach, a KBE engine has been developed and placed at the workbench of the designer. As the designer starts drawing a tube section, the KBE engine checks (online) the angles drawn by the designer against the manufacturing parameters that are based upon the capability of the bend machines and signals to him/her the necessary corrections to be implemented. The 3D tube leaves the CAD workstation only when it is ready to be manufactured. The tube data are filled and exported to the bending machine that finishes up the tube. The design-manufacture

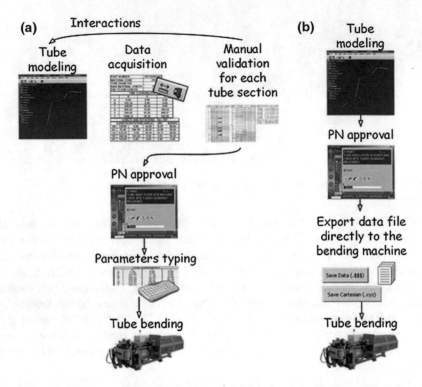

Fig. 2.17 Serial (**a**) versus integrated (**b**) tube design and manufacturing based upon the KBE engine

loop from the previous process has been replaced by a virtual and efficient dialogue between the designer and the manufacture engineer based upon the KBE engine.

It's worth noticing the differences and similarities between the DbF and KBE: a CAD environment is indispensable for both; the interaction between design and manufacturing is executed by the designer in DbF while it is rule based in KBE, without the designer interference. The majority of IPD tools, though, do not require a computer environment in order to promote the required integration.

2.6.3 Design for Excellence (DFX)

A great number of design tools are available to promote the integration of the technical areas of the product development, such as Design for Manufacturing (DFM),

Design for Assembly (DFA), Design for Recycling (DFR), Design for Service (DFS), Design for Packing (DFP), Design for E-Business (DFEB), Design for Automation (DFAut), and so forth.

All these DFX (Design for X or eXcellence, where X can be thought as a variable that can undertake the "values" M, A, R, S, P) have in common the aim to integrate the requirements of the technical area X into the conceptual design phase of the product. [3] The DFX design tools are indeed tools to be used by the IPD team members to advocate the best design option for the product regarding their technical areas. Among all possible product design options, the manufacturing area, through DFM, will point out those that best fit the manufacture requirements. Among all possible product design options, the maintenance area, through DFS, will point out those that best fit the maintenance requirements. It is quite possible that the DFM product option conflicts with that of DFS, thus raising an engineering tradeoff whose solution might partially fulfill both areas.

Some DFX IPD design tools are already well known and consolidated such as Design for Manufacturing (DFM), Design for Assembly (DFA) and Design for Service/Maintenance (DFS). Others are proposals yet to be tested such as Design for E-Business, Design for Nationalization, Design for Patent. All of them are related to some phase of the product life cycle and have one characteristic in common: an attempt to integrate the requirements of their product life cycle phase into the conceptual design phase of the product development process. It's worth stressing the words "attempt to integrate" because all the representatives of product life cycle phases will try to do the same—to advocate their cause. If just one phase or technical area prevails, the final configuration of the product would resemble of those shown in Fig. 2.18.

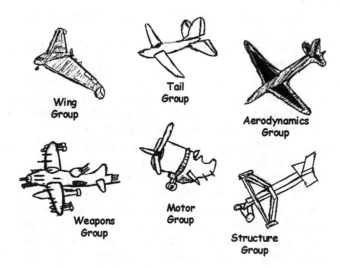

Fig. 2.18 Nonintegrated product development

In order to avoid this situation, the technical coordinator of the IPD team has to assure that all the requirements are taken into account, discussed, and incorporated into the product design in a balanced manner. It is correct to say then that in an IPD design team meeting, nobody leaves it 100 % happy and nobody does it 100 % unhappy.

Some DFX tools are implemented through guidelines which are derived from the merge of the design experience with the experience a person or a team has had in an "X" area. Some DFX guidelines are implemented through a systematic approach or method. One of them is the Design for Assembly (DFA). The guideline "Design for minimum part count" has been converted into a method by Boothroyd and Dewhurst in 1981 [4] and has evolved since then.

The Boothroyd and Dewhurst method for finding the minimum number of parts a product must have is based upon three questions the designer or team has to answer for each part or component which belongs to the product structure.

1. Does the component move relative to all other components already assembled?
2. Must the material of the component be different from those of the other components already assembled?
3. Must the component be separate from the other components already assembled to give access or disassembly them?

It should be noted that the causes for answering "Yes" to questions (1) and (2) must be related to the product's functionalities. A movement of a screw when it is been screwed receives a "No" answer for question (1). However, a power screw from a press has "Yes" as answer for the same question. Question (2) takes "Yes" for an answer whenever the component is used as electrical, thermal, or acoustic isolation, for instance.

Based upon the three questions, one must conclude that the minimum number of components or parts for *any* product is:

Minimum no. parts = No. of parts which has at least one "Yes" irrespective of the question + 1 (the prime or base part).

Figure 2.19 shows one product assembly example, before applying the method. The total number of components of this product is 20: four main components and 16 components for assembling the main components into the final shape. Also, the axle shown moves relative to the base due to the gear rotation.

If you go through the three previous questions, you conclude that the minimum number of components is two as shown in Fig. 2.19a.

The main benefit of the DFA analysis is not to determine the minimum number of components but rather, to search for new design proposals for the product that take into account the minimum number of parts. Suppose the design team has proposed the product design shown in Fig. 2.20a. Probably, the designer does not have the necessary information about the stamping process needed to manufacture the proposed design. Then it is sensible to think that he/she will ask the stamping specialist if the product "as designed" could be transformed into an "as built" product. By doing so, the DFA analysis fulfills its main objective, i.e. to promote the integration between design and manufacturing/assembly.

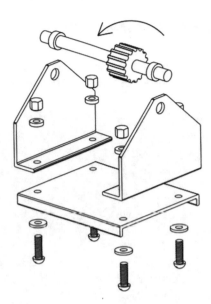

Fig. 2.19   An example of a product assembly

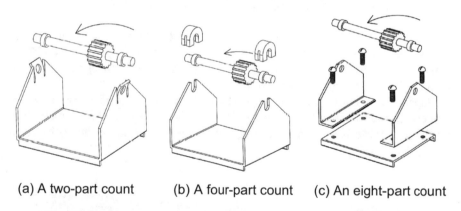

(a) A two-part count   (b) A four-part count   (c) An eight-part count

Fig. 2.20   Product redesign after the DFA analysis. **a** A two-part count **b** A four-part count **c** An eight-part count

The outcome of this design-manufacturing meeting could have been: our company hasn't have the necessary press to stamp the shape of the base or the volume of sales of the product does not justify the investment in a more complex stamping die.

The dialogue goes on. What about the design shown in Fig. 2.20b? It does not meet the minimum of part criterion but this is not the main issue. However, the stamping specialist still finds the stamping die rather complex and asks for an alternative product design. Finally, the design option depicted in Fig. 2.20c is the one that meets–partially–the design and manufacturing requirements.

2.6.4 Integrative Design Variables (IDV)

Costs, weight, center of gravity, and net electric power are examples of integrative design variables. The characteristics of these variables are the following:

1. There is a **target value** associated with them within a specific product development. Examples: the cost of an aircraft cannot be greater than \$14.5 M; the maximum weight of a robot end effector is 80 kg; the net power of a satellite is 2300 W.
2. These variables are **affected by almost all design decisions**. Examples: the choice of a single component impacts cost, weight, center of gravity, and perhaps net electric power if the component requires it for operation.
3. It is **easy to grasp the concept** around integrative design variables. Design people do not need to be lectured about them as their understanding is quite straightforward. Examples of design variables that do not meet this characteristic of IDV are: aerodynamic drag, wear, and stiffness.

The IDP design tools that address the integrative design variables are named DTX—Design *to* X rather than DFX—Design *for* X. Design to Cost (DTC), Design to Weight (DTW) [5], Design to Net Power (DTNP), Design to Center of Gravity (DTCG) are examples of the former.

Suppose that the IDV are displayed together within a specific product development and are regularly updated by the methods related to each of them. The resulting scenario is a technical managerial cockpit shown in an illustrative form in Fig. 2.21.

Fig. 2.21 Technical management cockpit

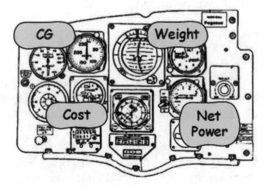

2.6.4.1 Design to Cost (DTC)—an Introduction to Integrative Variables

The basic equation that drives DTC is the following:

$$TC = TSP - TP \tag{2.1}$$

where:

- TC = Target Cost of a product
- TSP = Target Sale Price of a product
- TP = Target Profit of a company

The TSP variable is the starting point for DTC and it is usually obtained by the market intelligence department of a firm. TP is a firm internal variable that is usually set by its stakeholders. Then a product to be sold cannot cost more than TC.

DTC is also based on a process consisting of the following steps:

Step 1: Establish the product requirements. The design team may use some well-known design tools or methods to accomplish this step such as the Objective Tree method. In this method, a preliminary generic need is unfolded in several levels up to a stage where the need is converted into more meaningful and precise statements or requirements.

Step 2: Define the functional structure of the product. A design method that might help the design team work in this step is the Functional Analysis method. Similar to the previous method, Function Analysis is based upon a deployment activity, an overall, "black box"-like function is deployed in sub functions. The black box is transformed into a "transparent box."

Step 3: Elaborate design alternatives for the product. The Morphological Chart can be used to carry out this step.

Step 4: Estimate the cost of the functions. DTC prescribes the comparison the cost of the functions rather than the cost of the whole product. In doing so, the design team has more strict control over the design decision to meet the target cost of the product. The estimate cost of the functions is obtained from the matrix shown in Fig. 2.22.

The elements required for filling in the rows and columns of the matrix are the results from steps 2 and 3, respectively. Additionally, the design team has to search for the cost of the components required to fulfill the functions as indicated in the last line of the matrix. The remaining variables of the matrix are defined as follows.

V_{ij} = binary variable that indicates whether there exists a relationship between the component A_j and the function F_i ($V_{ij} = 1$) or not ($V_{ij} = 0$)

a_{ij} = variable that indicates—percentage—how much the component A_j influences the performance of the function F_i

Z_{ij} = partial cost of the function F_i with regards to the component A_j obtained as:

$$Z_{ij} = V_{ij} \times a_{ij} \times C_{Aj} \tag{2.2}$$

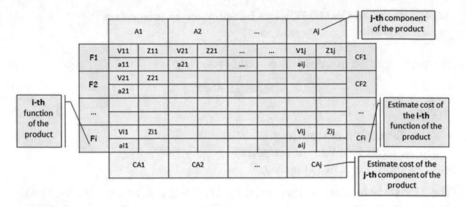

Fig. 2.22 Estimate function cost matrix

Finally, the estimate cost of the function F_i is calculated as presented in Eq. 2.3:

$$C_{Fi} = \sum_{j=1}^{k} Z_{ij} \qquad (2.3)$$

Equation 2.3 must be repeated for all **k** functions present in the functional structure of the product.

Step 5: **Compare the Estimate Cost of a Function (ECF) with the Target Cost of a Function (TCF)**. In order to accomplish this step, the design team has to figure out the target cost of the functions based on the target cost of the product. This can be achieved through the Value Analysis Method, where a survey must be carried out with the customers of a given product to determine how valuable the functions are to them. The results could be in a qualitative rank such as Low, Medium, or High.

Two possible results might come out from the comparison between the two categories of function costs:

$$ECF < TCF \qquad (2.4)$$

or

$$ECF > TCF \qquad (2.5)$$

The result presented by Eq. 2.4 is favorable for the design team and some actions could be derived from that so as to improve the components quality of this function or add some complementary sub functions to the main function. However, the technical coordinator of the IPD must be aware of what function could be causing the result from Eq. 2.5. Some actions that could be driven from that are to withdraw some sub functions or replace the components with lower cost counterparts.

Step 6: **Optimize the conceptual design of the product**. Instead of reasoning in an isolated manner, looking at the functions individually, the IPD design team could try to balance out the results from Eqs. 2.4 and 2.5—the surplus of one function could be used to rescue the deficit of another function. The scenario just described is an example of an IPD meeting agenda where one specific technical area, for instance, structures, meets another area, such as interiors, to sort out the best possible balanced solution that meets the design needs of as well the function target cost of both areas.

If you are considering an IDV different from cost, the logic from the previous steps remain the same, only substituting the measured variables.

 2.7 A Practical View

A very short list of IPD tools have been shown and discussed. Those IPD tools not presented herein are not less important than those tools presented herein. Integrated design methods such as QFD—Quality Function Deployment [6, 7], Design for Environment [8]; Design for Service [9] have to be applied to the product design so that all technical areas of the product life cycle are represented and heard at the conceptual design phase of the product development process.

Keep in mind that the final result of any DFX or DTX technique is the product. Take, for instance, an academic assessment as the product. Suppose you need to prepare an assessment about the subject Lean Product Development (LPD). The ordinary way to prepare it is to quickly define open questions such as "Discuss about the impact of LPD over the ISO 9000 certified companies." It's easy to think of nine more questions similar to that. However, the whole assessment process includes the correction and marking of the assessments. That can take a lot of time that is directly proportional to the number of students.

Design for Correction is an application of the DFX techniques to academic or technical assessments [10]. The assessment (that's the product) takes into account the requirements of the correction process as well. Naturally it takes longer to elaborate upon the questions compared to the traditional way, but the whole assessment process is shorter because the correction can be done as shown below.

The DFC questions are prepared in a way that requires the students to establish the relationship among a number of concepts, approaches, and techniques discussed during the course. The students are free to check their class notes, slides, books, and papers. A sample question is shown below.

Mark the correct alternative(s) with regards to Lean Development Product (LPD):

(a) It is more important to understand how the Lean philosophy is applied to the Product Development Process (PDP) than to know the lean techniques and tools.
(b) Because LPD radically differs from Integrated Product Development (IPD), it makes LPD a very difficult matter to be understood by western companies.
(c) As Knowledge Management (KM) is a weak characteristic of LPD, the two subjects are complementary and create a sustainable and competitive advantage for the companies.
(d) The continuous improvement associated with LPD has little impact over the PDP performance indicators once the majority of the companies are already ISO 9000 certified.
(e) Based upon the triad Knowledge, Skills and Attitudes (KSA), the teaching of LPD in western companies has to be focused on Knowledge.

The open question "Discuss about the impact of LPD over the ISO 9000 certified companies," is replaced by five alternatives with more strict content. Nevertheless, in all the alternatives the student has to review several concepts and the relationship among them. In the sample question, the concepts of LPD philosophy, LPD tools, IPD, KM, KSA are intentionally mixed.

The test lasts 60 min and the students keep their test sheets for correction. The lecturer starts the oral correction by stating the correct answer for each question (answer "A" on the sample question). A student might argue that another answer is also correct; having to explain what sustains his/her choice. Other students might join the discussion and turn the correction process into a "Greek Agora Square." Naturally, the lecturer has to keep the discussion under control, avoiding the corporatism syndrome. Eventually, the argumentation of the student might be taken into account and the lecturer would consider the student's choice correct. It is not the case, though, for the presented sample question.

References

1. Andreasen MM, Hein L (1987) Integrated product development. IFS-Springer Verlag, Berlin
2. Paris C, Salas E, Cannon-Bowers J (2000) Teamwork in multi-person systems: a review and analysis. Ergonomics 43(8):1052–1107
3. Huang GC (1996) Design for X: concurrent engineering imperatives. Chapman&Hall, London
4. Boothroyd G, Dewhurst P (2010) Product design for manufacture and assembly, 3rd edn. CRC Press, New York
5. Furtado LF, Villani E, Trabasso LG, Silva CEO (2013) DTW: a method for designing robot end-effectors. J Braz Soc Mech Sci Eng. doi:10.1007/s40430-013-0109-8
6. Akao Y (1995) Quality function deployment: integrating customer requirements into product design. Productivity Press Inc., New York

7. Araújo MF, Trabasso LG (2013) Applying QFD to business development environment. J Braz Soc Mech Sci Eng 35:1–14. doi:10.1007/s40430-013-0010-5
8. Guerato AM, Araújo CS, Trabasso LG (2010) A method to obtain context based DFE criteria lists applicable to the development of friendly recycling composite aeronautic structures. Product: Management & Development (IGDP), v.8, pp 71–79
9. Vezzoli C, Kohtala C, Srinivasan A (2014) Product-service system design for sustainability. Greenleaf Publishing
10. Pessôa MVP, Trabasso LG (2014) A Lean way to teach Lean Product Development to Graduates. In: IIE Engineering Lean and Six Sigma Conference and the Lean Educator Conference: Orlando, USA

Chapter 3
Lean Thinking

Applying lean to the PDP takes IPD a step further to create Lean Product Design and Development (LDP). While LPD keeps the integrative aspect of IPD, the LPD is rooted in the lean philosophy, which advocates full commitment to delivering the pulled value (do the right), and yet continuously works on eliminating the waste from the PD process (do right). This idea of "doing more with less", which was the after war reality in Japan, is nowadays commonsense everywhere. Environmental and scarce resources concerns pull the modern companies to a more "lean philosophy aligned" behavior. Considering that the LPD is this book's main subject, which will be discussed from now on, this chapter focuses on presenting a brief review of the lean philosophy and its impact on the manufacturing and on the product development systems.

3.1 Introduction

In the 1950s, Eiji Toyoda, Shigeo Shingo and Taiichi Ohno at Toyota Motor Company in Japan, developed the Toyota Production System (TPS). The TPS, which most people now associate with the term *Lean*, or more with the *Just-in-Time* (JIT) principle, was born when the Japanese car industry was stuck in a severe crisis. At that time, it became clear that the only way to escape from the possible impending doom of this industry were drastic changes in efficiency and productivity. This change happened through lean thinking (or philosophy) that means thoroughly working on waste reduction while guaranteeing the value creation. The lean philosophy was presented to the rest of the world by the results of the MIT International Motor Vehicle Program (IMVP) whose goal was to compare the performance differences between car companies operating with traditional mass manufacturing systems and those using the TPS [1]. Besides seeking an optimum flow (waste reduction), lean philosophy focuses on value identification and value delivery to the customer.

© Springer International Publishing AG 2017

M.V.P. Pessôa and L.G. Trabasso, *The Lean Product Design and Development Journey*, DOI 10.1007/978-3-319-46792-4_3

The lean thinking or philosophy main's goal is to getting closer and closer to provide what the customers exactly want: the value pulled by them. In order to do so a simple and yet difficult set of principles shall be followed [2]

1. **Specify value**: The value, as defined by the final client, is the basis of lean thinking and guide all processes in the company. Without identifying the value one cannot discern value added activities from wasteful activities.
2. **Identify the value stream**: The value stream is a theoretical and ideal sequence of exclusively value-added tasks, where a value-added activity transforms the deliverables of the project in such a way that the customer recognizes the transformation and is willing to pay for it. By lining up the value delivery activities the wasteful activities are reduced if not eliminated, therefore more time, money, human resources, etc. can be redirected to what really matters.
3. **Guarantee the flow:** All the value-added activities should be conducted without interruption.
4. **Pull the value**: No activity in the value stream should be produced without being requested by the next activity in the flow.
5. **Seek perfection**: The relentless continuous improvement is the motor that sustains and evolves the lean philosophy.

While the first two principles guarantee the value delivery, the remaining three work on waste reduction. Indeed, seeking perfection, which is the reason for continuous improvement, means relentless continuous waste reduction.

As a target condition, any Lean System should deliver maximum value, while reducing waste. The term "waste reduction," though, is not limited to waste itself, but also includes unevenness and overburden (Fig. 3.1) where:

- **Value**: Value to a stakeholder is the total and balanced perception from all the benefits provided by the results of the life cycle processes. "Total perception" is considered to be not only results directly related to the product, but also expectancies related to cost, on time delivery, risk level, etc. [3].
- **Waste** (Muda, 無駄 or ムダ): Waste refers to all elements of a process that only increase cost without adding value or any human activity that absorbs resources but creates no value; any activities that lengthen lead times and add extra cost to the product for which the customer is unwilling to pay. [2, 4].

Fig. 3.1 Lean system target condition

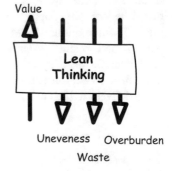

- **Overburden** (Muri, 無理): Overburden is viewed as pushing a machine, process, or person beyond natural limits [4].
- **Unevenness** (Mura, 斑 or ムラ): The results from an irregular production schedule or fluctuating product volumes caused by internal problems create unevenness [4].

Waste can manifest as excess inventory, excessive production, extraneous processing steps waiting, unnecessary movement, and defective products, etc. All these "waste" drivers intertwine with each other to create more waste [5, 6], eventually impacting beyond the specific area where their root causes appear and escalating even to the management of the corporation itself.

The lean thinking success, though, is not limited to manufacturing. It can be applied to other processes with high cost reduction and quality improvement potential, which is the case of product development. In fact, the success of Toyota relies on the application of lean philosophy on the product development rather than on the manufacturing.

By teaching Lean Product Design and Development through the years we could see how the students and practitioners made connections about applying the lean philosophy into the most diverse areas, including the daily life.

The transitioning to lean is a mindset change; it's a cultural change in the organizations and an attitude change in each individual. Therefore the lean thinking person takes the philosophy to all aspects in his/her life.

3.2 Lean Manufacturing

The Toyota Production System (TPS), sometimes called *lean manufacturing system* or a *Just-in-Time* (JIT) system, is a way of "making things" by applying the lean philosophy into a company, and which has evolved into a world-renowned production system.

The TPS imbues all aspects of production in pursuit of the most efficient methods. The TPS has evolved through many years of trial and error to improve efficiency based on the Just-in-Time concept, where the ideal conditions for making things are created when machines, facilities, and people work together as needed to add value without generating waste [7].

Figure 3.2 shows the TPS house [8]. The house metaphor depicts all the TPS elements. Therefore, in order to sustain the roof, its two pillars (the TPS objectives, jidoka and just-in-time) have to be sustained by a solid base of standards and principles. The motivated team populates this house, and is the key element to set the TPS in motion [9].

These principles have been adopted by diverse sectors of industry such as aerospace, consumer products, metal processing and industrial products.

The TPS house shows that the lean system is a complex socio-technical system and not only a set of lean labeled tools and techniques. Therefore, having a

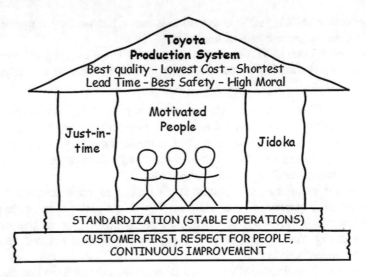

Fig. 3.2 The Toyota production system house

true lean manufacturing system requires changes and adaptations in several dimensions: attitude, standards, tools, techniques, people, etc.

Next we discuss the TPS's two main pillars: just-in-time and *jidoka*. In our experience, the concepts behind these two pillars remains the same whatever is the process or area you are going to apply the lean philosophy.

3.2.1 Just-in-Time[1]

Just-in-Time means making only what is needed, when it is needed, and in the amount needed; this can eliminate waste, inconsistencies, and unreasonable requirements, resulting in improved productivity and continuous flow.

In order to deliver a product (a vehicle, in the case of Toyota) ordered by a customer as quickly as possible, the product is efficiently built within the shortest possible period of time by adhering to the following:

- When a product order is received, a production instruction must be issued to the beginning of the vehicle production line as soon as possible.
- The assembly line must be stocked with the required number of all needed parts so that any type of ordered product can be assembled.
- The assembly line must replace the parts used by retrieving the same number of parts from the parts-producing process (the preceding process).

---

[1]Adapted from [9].

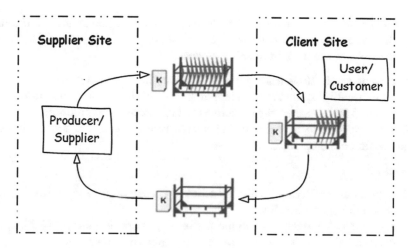

Fig. 3.3 Conceptual diagram of the Kanban system

- The preceding process must be stocked with small numbers of all types of parts and produce only the numbers of parts that were retrieved by an operator from the next process.

Whatever the steps are of your production process, the logic remains the same—the next process pulls what it needs from the previous one. In the TPS, a *kanban system* is used to control the flow. The *kanban* system has also been called the *Supermarket method* since it is inspired by the product control cards used in these stores. These cards normally contain product-related information, such as a product's name, code, and storage location. At Toyota, when a process refers to a preceding process to retrieve parts, it uses a *kanban* to communicate which parts have been used.

By having the next process (the customer) go to the preceding process (the supermarket) to retrieve the necessary parts when they are needed and in the amount needed, it was possible to improve upon the existing inefficient production system. No longer were the preceding processes making excess parts and delivering them to the next process (Fig. 3.3).

The just-in-time pillar acts on guaranteeing an effective and flexible process and reducing process waste, work in process, and inventory. In this kind of process waste is easier to see, and improvements are easier to implement.

3.2.2 Jidoka[2]

For the Just-in-Time system to function, all of the parts that are made and supplied must meet predetermined quality standards. This is achieved through *jidoka*. The objective of jidoka is doing it right the first time (i.e.,"first time right").

---

[2]Adapted from [9].

Jidoka can be loosely translated as "automation with a human touch," meaning that when a problem occurs, the equipment should be stopped immediately, preventing defective products from being produced:

- A machine safely stops when the normal processing is completed.
- Should a quality/equipment problem arise, the machine detects the problem on its own and stops, preventing defective products from being produced.
- This is opposed to a machine that simply moves under the monitoring and supervision of an operator.

As a result:

- Only products satisfying quality standards are passed on to the following processes on the production line.
- Operators can confidently continue performing work at another machine, and work on identifying the problem's cause to prevent its recurrence.

Jidoka supports the defect-free production line vision. As a consequence defects are not tolerated, and whenever a defect is spotted the process should guarantee that no defective parts are produced until the problem's cause is identified and solved.

3.3 Lean Product Development

In PDP, lean thinking goes beyond systematic waste reduction and the application of lean manufacturing techniques to product development. To allow Lean Product Development, besides lean itself, the project plan must allow value creation while providing for waste reduction.

In PD, adding customer value can be less a function of doing the right activities (or of not doing the wrong ones) than of getting the right information in the right place at the right time. Hence, the focus of lean must not be restricted to activity "liposuction" (waste reduction), but must address the PD process as a system (value creation). To guarantee the value creation and to create the needed countermeasures against waste, we adapted the five lean principles initially proposed by Womack and Jones [2] to the PD context:

1. **Specify value**: In a program or project, the value is the *raison d'être* of the project team, which means they must understand all the required product/service characteristics regarding the value that all stakeholders of the program expect to receive during the product life cycle.
2. **Identify the value stream**: Consequently, the Product Development Process must be simple, highlighting key dates and responsibilities and defining optimized information flows (what, when, sender, receiver, and media) in order to prevent excessive data traffic and promote efficient communication.

3. **Guarantee the flow**: The ideal PD process should work congruently with the single-piece flow in manufacturing, representing a value flow from conception to production, without stops due to bureaucracy and loop backs to correct errors. Every development value flow obstacle (functional departments, executive gate meetings, firefighting, changing requirements, management interference, etc.) must be eliminated.
4. **Pull the value**: Instead of pushing scheduled activities, which themselves push information and materials through the development process, pull events must be defined. Different from tall gates, where information batches are created, pull events guarantee the value flow, make quality problems visible, and create knowledge.
5. **Seek perfection**: The continuous improvement of the development process is achieved by the capability of the process and effective knowledge management. This knowledge is systematically documented and disseminated through trade curves which everyone can access and is expected to use, including management. The relentless continuous improvement is the motor that sustains and evolves the lean philosophy.

Lean Development, although at first glance looks a lot like the Integrated Product Development since it includes the same practices, proposes a more organic view of the process. In this way, the value flow is sustained by two pillars (Fig. 3.4), based on waste reduction (efficiency: do the job right) and the creation of value (efficacy: do the right job).

This vision must be achieved through maximum simplification of the process (removing the activities that do not add value), and enforcing the activities of prototyping and testing; the idea is to maximize experimentation and learning. The development project manager must assume roles beyond coordinating and motivating, he/she is expected to coach the engineers and technicians under his/her supervision, in a constant quest for innovation. At this point, he/she is concerned with the enhancement of organizational learning and knowledge management.

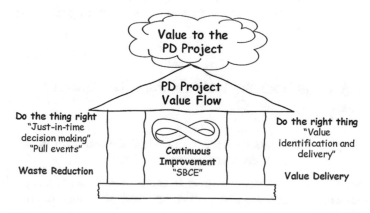

Fig. 3.4 Value creation pillars to product development

In the integrated development approaches, this aspect from working in teams was not valued.

Indeed, the application of Lean Philosophy has three core elements to the PDS:

1. **Do the right thing** = deliver the Value = create the correct products: Create product families and projects that create value for all stakeholders of the enterprise architectures. The understanding of value in the PDP context is paramount (see Chap. 4), and correct value identification is the most critical success factor in any PD project (see Chap. 10), once no one is going to buy and/or use the wrong product.
2. **Do the thing right** = use effective engineering processes: Eliminate waste (for a complete description of the PD wastes see Chap. 5) and improve cycle time and quality engineering while achieving effective integration between the development process and the company using lean engineering to create value on the interfaces between the development process and the various parts of the company. Just-in-time decision making (JIT-DM) supports waste reduction by guaranteeing that decision is only taken (and is taken quickly) when the necessary information is available. To operationalize JIT-DM we recommend planning you development by defining pull events, which pull the necessary information from the development team at the right moment during the PD project (see Chap. 12).
3. **Never rest on previous successes** = continuous improvement: Be a learning organization and keep improving every day. One can improve both the executing process and its deliverables (Chap. 6). The PDP goal is to gradually reduce the risk and uncertainty while develop a new product (Chap. 1). As a consequence, the PDP itself functions a continuous improvement process. The Set-Based Concurrent Engineering (SBCE) , which is further detailed in Chaps. 9 and 11, plays a key role to guarantee the continuous development results evolution towards delivering the identified pulled value.

We can see (Fig. 3.5) the elements of Lean Thinking applied in the Lean Wheel System (value, waste, continuous improvement); they make the wheel hub which sustains the remaining parts of the system to be further described in Part II of this book.

3.4 Key Aspects from the Lean Product Development System

The wheel and tire elements from the "Lean Wheel System" (Fig. 3.5) encompass the key aspects of the Lean Product Development System:

1. **Culture**: Culture guarantees the creation and maintenance of the organizational environment; it gives the necessary "code of conduct" to the people performing the Lean PDP.

Fig. 3.5 Lean wheel system
elements

2. **Knowledge management**: In order to keep moving forward, the company has
   to be a learning organization; failing to learn may result in repeating the same
   mistakes and reinventing.
3. **Organizational structure**: Hierarchical structure defines the company and the
   peoples' roles necessary to perform the Lean PDP.
4. **Tools and Technology Subsystem**: This encompasses all the tools and
   Technologies used by the development organization, CAD systems (*Computer
   Aided Design*), digital manufacturing, test technologies, etc.
5. **Lean Development Process**: The process includes all activities required to
   bring the product from concept to production, passing through the entire value
   stream. It focuses on both identifying the value and guaranteeing the flow and
   the emergence of the final product.

These elements are going to be further described in Parts III and IV of this book.
The list on Table 3.1 summarizes the philosophy's impact through the Lean
Product Development Organization.

 3.5 A Practical View

Many people get lost while trying to apply the Lean Philosophy. This fact is not
limited to Lean Product Development, but occurs in any process improvement
effort. They get lost because they work hard on understanding how Toyota works
and they start to believe that being lean is replicating Toyota (Fig. 3.6).

Table 3.1 Key elements of the LPDS [2, 10]

Element	Description
Culture	• Support excellence and relentless improvement • Adapt technology to fit your people and process • Align your organization through simple, visual communication
Knowledge management	• Standardized "performance tradeoff" data are collected for each alternative • Use powerful tools for standardization and organizational learning • Engineers are required to be knowledgeable about all solutions • Detailed engineering checklists and design standards are used to assure focus on product performance • Fully integrate suppliers into the product development system • Build in learning and continuous improvement
Organizational structure	• Managers are the most technically competent in engineering: "your boss can always do your job better than you" • The manager's primary role is to teach by assigning questions (mentoring) • Authority and rewarding in the system derives from technical knowledge and competence • Develop a value-centered system to integrate development from start to finish • Organize to balance functional expertise and cross-functional integration
Process	• No elaborate sub-schedules; chief engineer sets "key integration events" • Work is pulled to these events • Milestones are never missed • Multiple alternatives are developed for each subsystem • Combinations that meet performance tradeoffs "survive" • Establish customer-defined value to separate value-added from waste • Create leveled product development process flow • Utilize rigorous standardization to reduce variation, and create flexibility and predictable outcomes
Tools and technology	• The lean tools and technology are those you use in the lean way, not the "lean labeled tools" • The tools and techniques do not make you lean, the way you use the tools is what makes them lean

Wrong! In order to adopt the Lean Philosophy one has to understand its roots and adapt. By studying Toyota you can learn how they created/adapted tools, techniques, and process into their own reality, in order to apply the philosophy. Whenever you see a successful approach performed by any company, you must ask what are they achieving and why did they work that way.

The lean philosophy has its roots in the post-war Japan. At that time the western way of designing, developing, and producing cars were the best practices to be copied. Toyota managers understood how their company, culture, and country differed from the western benchmark; they studied and positively exploited these differences. There is no reason to not believe that you and your company might do the same and become the next benchmark.

Fig. 3.6 No two flowers are
the same

A good way to do so is applying the lean product design and development
method presented in this book, having your product design development process (or any process you want to improve) as the "product" to be designed and
developed.

References

1. Womack JP, Jones DT, Ross D (1990) The machine that changed the world. Rawson
   Associates, New York
2. Womack JP, Jones DT (2003) Lean thinking. Free Press, New York
3. Pessôa MVP (2006) Proposta de um método para planejamento de desenvolvimento enxuto
   de produtos de engenharia (Doctorate Thesis) Instituto Tecnológico de Aeronáutica: São José
   dos Campos
4. Morgan JM, Liker JK (2006) The Toyota product development system. Productivity Press,
   New York
5. Pessôa MVP, Seering, W, Rebentisch E, Bauch C (2009) Understanding the waste net: a
   method for waste elimination prioritization in product development. In: Chou S et al. (Org.)
   Global perspective for competitive enterprise, economy and ecology. Springer-Verlag,
   London, pp 233–242
6. Pessôa, MVP, Seering W (2014) Trapped on the waste net: a method for identifying and prioritizing the causes of a corporation's low product development performance. In: Marjanovic
   D et al (ed) Proceedings of the design 2014 conference, vol 3. The Design Society, Glasgow,
   pp 1641–1650
7. Rother M (2010) Toyota Kata: managing people for improvement, adaptiveness and superior
   results. McGraw-Hill, New York
8. Liker JK (2004) The Toyota way: 14 management principles from the world's greatest manufacturer. McGraw-Hill, New York
9. www.toyota-global.com/company/vision_philosophy/toyota_production_system/
10. Cleveland J (2006) Toyota's other system -this one for product development. Automot Des
    Prod 118(2):18–22

Part II
The Wheel Hub

Part II includes the "Core Lean" elements of value creation, waste reduction, and continuous improvement (Fig. 1). Chapter 4 presents the concept of value applied to the Lean Product Development System. In sequence, Chap. 5 analyzes the concept of waste in the Product Development Process. Finally, Chap. 6 discusses how continuous improvement and adaptation allows the emergence of the true Lean Organization.

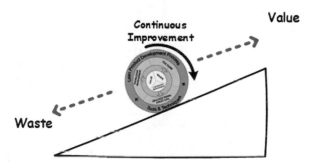

Fig. 1 Wheel hub elements

Chapter 4
Value on Product Development

The term value is rather ambiguous, even in the product development field. This chapter presents how project management, value engineering and lean product development define value and highlights the consequences of these different understandings. We stress that, in lean product design and development, the goal is to understand the value pulled by all stakeholders in the value chain. Although fulfilling the needs from the external stakeholders (user, customer, shareholder/sponsor, etc.) have more impact in product success (do the right product), fulfilling the internal stakeholders' needs (suppliers, design and development team, production, development partners etc.) guarantee a smoother PDP execution (do the product right). Considering the lean two pillars, we address the jidoka by understanding the value pulled by the external stakeholders, while the just-in-time refers to the internal stakeholders' needs.

4.1 Introduction

Lean Development is based on the value to the stakeholders. The word "value," though, is applied to several areas and with diverse meanings. According to the *American Heritage® Dictionary of the English Language* [1] some of its meanings are:

- An amount, as of goods, services, or money, considered to be a fair and suitable equivalent for something else; a fair price or return.
- Monetary or material worth: the fluctuating value of gold and silver.
- Worth in usefulness or importance to the possessor; utility or merit: the value of an education.
- A principle, standard, or quality considered worthwhile or desirable.
- Precise meaning or import, as of a word.
- Mathematics: An assigned or calculated numerical quantity.
- Music: The relative duration of a tone or rest.
- The relative darkness or lightness of a color.

© Springer International Publishing AG 2017
M.V.P. Pessôa and L.G. Trabasso, *The Lean Product Design and Development Journey*, DOI 10.1007/978-3-319-46792-4_4

- Linguistics: The sound quality of a letter or diphthong.
- One of a series of specified values: issued a stamp of new value.

As a consequence, the understanding of what is "value" is not uniform in the literature: (1) in the management of traditional projects, the value is a consequence of the execution of activities and generation of results, so greater efficiency should create more value; (2) in engineering, value is a function of the obtained benefit by its related cost; and (3) in lean philosophy, the value for a particular stakeholder is the sum of all benefits perceived by him, through the development results, which, in addition to the final product, includes all the intermediate results, the use of which composes this experience.

Considering the importance of the value concept to Lean Philosophy, and the different meanings from different approaches related to product development, this chapter will further explain the point of view from the disciplines of Project Management, Value Engineering, and Lean Product Development.

4.2 "Value" and Project Management

From the Project Management point of view, value is defined in terms of scope, budget and schedule. This fact is evidenced by the Earned Value Analysis, EVA,[1] which is a method for performance measurement widely used by project management practitioners.

The EVA requires the creation of an integrated project baseline against which performance can be measured during the duration of the project. The EVA can be applied to all projects and in any industry; it works by monitoring three dimensions (Fig. 4.1):

1. **Planned Value (PV)**, or Budgeted Cost of Work Scheduled (BCWS), is the assigned budget to any scheduled work to be accomplished during the project.
2. **Earned Value (EV)**, or Budgeted Cost of Work Performed (BCWP), is the value of the work performed expressed in terms of the approved budget assigned to that work as scheduled.
3. **Actual Cost (AC)**, or Actual Cost of Work Performed (ACWP), is the total cost actually incurred and recorded in accomplishing the scheduled work.

To determine PV, the project manager must ascertain (1) how much physical or intellectual work has been scheduled to be completed at a certain point in time, and (2) management's authorized budget for this authorized work. The planned value is simply the direct fallout of those detailed tasks specified on the project master schedule.

---

[1]For a complete EVA description, check Fleming & Koppelman [2], and PMI [3].

Fig. 4.1 Project performance and EVA

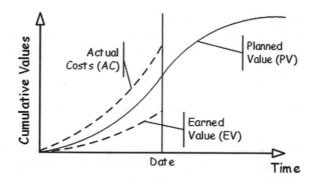

To measure EV for the same reporting period, the project team will need to determine (1) how much of the authorized work they actually accomplished, and (2) the amount of management's original budget for the accomplished work.

To determine the AC, the project team needs to know how much money was spent converting the PV into EV during the measurement period.

To perform the EVA four indicators are used:

1. **Cost Variance**: $CV = EV-AC$, where positive values represent good performance.
2. **Schedule Variance**: $SV = EV-PV$, where positive values indicate the anticipation of planned deliveries.
3. **Cost Performance Index**: $CPI = EV/AC$, where $CPI < 1$, $CPI = 1$ and $CPI > 1$ indicate over budget, on budget, and below the budget, respectively.
4. **Schedule Performance Index**: $SPI = EV/PV$, where $SPI < 1$, $SPI = 1$ and $SPI > 1$ indicate over time, on time, and behind the schedule, respectively.

The development project, represented in Fig. 4.1, is both over budget ($EV/AC < 1$) and behind the schedule ($EV/PV < 1$).

The EVA, though, is not suitable for measuring the value in the early stages of development, since much of the value recorded is reworked in later stages, without being able "disaggregate it" [4].

Additionally, the total time of an activity is different from the time of effective value added once we discern between (1) calendar time—the total duration of activity from start to finish; (2) working time—the percentage of calendar time in which resources are actually available and performing work; (3) time of adding value—the part time job that really is value added, i.e. it is something for which the customer is willing to pay. By this definition, not all working time necessarily add value. Even value-added activities do not add value during its complete calendar time. Thus there is a change of focus, where the value does not emerge simply from the efficient execution of activities (do the job properly), but depends on the results (do the job right): there is no reason to do the wrong work properly.

In fact, traditional project management is concerned with operational efficiency, where a project is considered successful if it is completed on time, on budget, and within specifications ($AC = EV = PV$, or $SPI = CPI = 1$). Although necessary, operational efficiency is not enough, it must be primarily meeting the needs of business that lead to the creation of the project (do the job right).

4.3 "Value" and Value Engineering

Value Engineering (VE) is an organized/systematic approach that analyzes the functions of systems, equipment, facilities, services, and supplies to ensure they achieve their essential functions at the lowest life-cycle cost consistent with required performance, reliability, quality, and safety. Typically, the implementation of the VE process increases performance, reliability, quality, safety, durability, effectiveness, or other desirable characteristics [5].

VE is a systematic method to improve the "value" of goods or products and services by using an examination of function. Value, as defined, is the ratio of function to cost. Value can therefore be increased by either improving the function or reducing the cost. It is a primary tenet of value engineering that basic functions be preserved and not be reduced as a consequence of pursuing value improvements.

Because "costs" are measurable, "cost reduction" is often thought of as the sole criterion for a VE application, and indeed, cost reduction is primarily addressed in this approach. It is important to recognize, however, that increased value is the real objective of VE, which may not result in an immediate cost reduction.

In fundamental terms, VE is an organized way of thinking or looking at an item or a process through a functional approach. It involves an objective appraisal of functions performed by parts, components, products, equipment, procedures, services and so on—anything that costs money. VE is performed to eliminate or modify any element that significantly contributes to the overall cost without adding commensurate value to the overall function.

Therefore, the sense of an item's value depends on your point of view:

- According to the seller: value = function/cost.
- According to the buyer: value = benefits/price.

Where the function can be defined as: (1) purpose of a product or system operating at its normally prescribed manner; (2) "something" that makes the item work or sell; and (3) objective of activities performed by one or more organizational units considered systems.

Value Engineering seeks to identify the primary and secondary functions of a product and set the various alternatives available for these functions in order to define the best product architecture based on the cost/benefit ratio.

VE is not primarily centered on a specific category of the physical sciences; it incorporates available technologies, as well as the principles of economics and business management, into its procedures. When viewed as a management discipline, it uses the total resources available to an organization to achieve broad management objectives. Thus, VE is a systematic and creative approach for attaining a return on investment by improving what the product or service does in relation to the money spent on it.

4.4 "Value" and Lean Product Development

The value, as defined by the customer and the product user, is the basis of lean thinking, thus the development provides no value unless it meets the expectations of these stakeholders [6–8]. This value, though, must be translated into measurable functions and non-functional parameters, which can be designed, produced, and verified.

In a project or product development program, identifying value means understanding the necessary characteristics of the product and/or service, and determining the value that the program stakeholders expect to receive throughout the product life cycle: the focus is on information flow which is usually non-linear and iterative.

The nature of the value created by each project team varies greatly from industry to industry. Providing an incorrect service or product to the customer means waste, even if the development and commercialization processes are efficient and error-free. The value must be systemic, where any function, feature, or characteristic from a specific product part or subsystem adds value only if perceived in this way by the customer or any stakeholder in the value chain [9].

The value identification, therefore, is a critical development success factor. Problems that go unnoticed at this stage are the most expensive to resolve since they cause more waste and the consequent rework.

Focusing only on the customer is a very simplistic strategy, particularly in industries of complex and highly aggregated value products. The failure to take into account all key stakeholders (both those who positively and negatively impact the development project) as well as to negotiate the value trade-off among them, can lead to an incorrect result (no value added). This entails the inability to deliver the promise, resulting in the consequent failure of the program.

Indeed, even though the customer/user is the primary and most important stakeholder, there are several other stakeholders, inside and outside the company, that either have expectations about the product or that might influence on how smooth the development project flows through the value chain (see next topic).

There is also the challenge of managing non-value-add "needs" pushed into the product, service, or result to be provided, and might cause the failure or stagnation of the development project, such as [10]:

- Preconceived solutions, which worked in the past and have been institutionalized as "monuments".
- The existence of a powerful advocate with a personal interest in a particular solution.

- The tendency to underestimate the difficulty in developing a new technology, especially if it occurs simultaneously with the development of a new product or system based on this technology. Sometimes, sophisticated technologies both exceed the budgets of customers and impede their real desires.

To specify the true value, one must often relinquish resources, technologies, and practices already used in the business because the current paradigm does not allow correctly recognizing the value [7]. This can occur when the company's focus is on the short-term where prevailing actions will lead to immediate returns, but which may become future losses.

Changes in the business environment are also key factors to be addressed in the value identification process. In the aerospace industry, for instance, rather than simply favoring performance, the expected received value throughout the life cycle and the related cost of ownership has gained importance. Thus, less tangible features such as convenience, reliability, and maintainability shall be considered. In addition, greater attention is given to the operating environment and the needed infrastructure to support its use in the long run [10].

Only the systematic identification of value leads to a correct product concept and value proposition.

During the PDP execution: (1) development activities generate results; (2) the results are delivered to relevant stakeholders; and (3) the results allow the perception of the benefits by these stakeholders. Furthermore, since the expected value for each stakeholder cannot be delivered only by the products, services, or final results, ensuring a complete and consistent value delivery throughout the product life cycle must be supported by both the development process itself and the performing organization.

Figure 4.2 confirms this position by showing the interdependence between these three dimensions. The product, while produced by the processes determines the remaining lifecycle processes. The organization performing the process is itself dependent on the procedures, and on the other hand, constrains these very

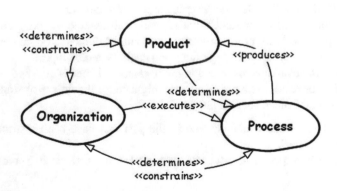

Fig. 4.2 Product, process, and organization

processes; since not every kind of process can be performed by a given organizational structure. Therefore, product and organization are interdependent [11].

> We define value for a given stakeholder as "the total and balanced perception, resulting from **the various benefits delivered through the product/process lifecycle."**

This total and balanced perception means not only that the product or service meets its functional expectations, but also all other perceivable dimensions. Thus, functional and non-functional aspects, as well as meeting the constraints of budget, schedule, and risk aversion are also considered in the overall perception of value since they relate to the expected benefits for each stakeholder (Fig. 4.3).

Stakeholders are individuals or organizations actively involved in the development, or whose interests may be affected by its execution or completion (with either success or failure). [3] Stakeholder identification is therefore crucial because they (1) are those who demand value; and (2) may have a positive or negative influence on the development success. The developing organization, in order to ensure the development's success, must identify their expectations and, to the extent possible, manage their influence and solve conflicts among their interests.

Deliverables are the outputs from the product lifecycle (which includes the PDP). A deliverable is tangible and has one or more specific recipients. Benefits, however, come from the deliverables used by the stakeholders, and may be tangible or intangible. For instance, the use of engineering design models and prototypes might generate benefits for the engine development team; while the engine itself will benefit other development teams (from other parts of the car), the end customer, etc.

Considering that the product design and development process has several stakeholders, both internal and external to the organization, the value for a given product development project is the sum of the value for each related individual stakeholders. The product development project value is the target condition or vision that the product development team has to achieve at the end of the project (Fig. 4.4).

While the externally pulled value is perceived only after the product launch, and guarantees its market success, the internally pulled value is perceived through the PDP's phases and guarantees the smooth PDP flow.

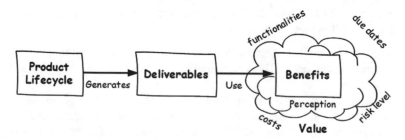

Fig. 4.3 Value perceptions

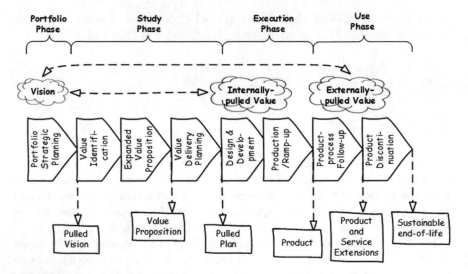

Fig. 4.4 Value and the product development process

4.5 Value Through the Value Chain

Understanding how value flows through the value chain and how the external stakeholders perceive it is paramount to a high performance product development. By understanding how value flows, particularly through the value chain's primary activities, important stakeholders and their pulled value are identified. By understanding how the external stakeholders perceive (and value) the company's value chain activities, an improved value chain can be also a result of the development project.

A value chain is a chain of activities for a firm operating in a specific industry. It models how businesses receive raw materials as input, add value to the raw materials through various processes, and sell finished products to customers [12]. Therefore, it comprises all the organization's primary and support activities, not forgetting all the interfacing activities with other organizations within the supply chain (Fig. 4.5).

Primary activities relate directly to the physical creation, sale, maintenance and support of a product or service. They are the activities that really add value. They consist of the following:

- **Inbound logistics**: These are all the processes related to receiving, storing, and distributing inputs internally. Your supplier relationships are a key factor in creating value here.
- **Operations**: These are the transformation activities that change inputs into the outputs that are sold to customers. Here, your operational systems create value.
- **Outbound logistics**: These activities deliver your product or service to your customer. These are things like collection, storage, and distribution systems, and

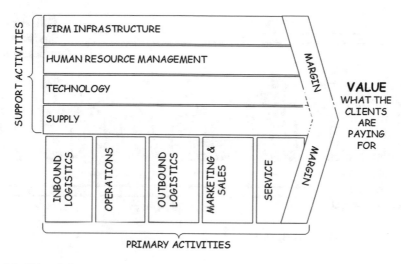

Fig. 4.5   Value chain

they may be internal or external to your organization. Your distribution channels are a key factor in creating value here.

- **Marketing and sales**: These are the processes you use to persuade clients to purchase from you instead of your competitors. The benefits you offer and how well you communicate them are sources of value here.
- **Service**: These are the activities related to maintaining the value of your product or service to your customers once it's been purchased. These services vary according to your strategy and product and might include training, maintenance, automatic resupply, etc.

Support activities, even though they not create value themselves, are necessary to support the value-added activities, aka primary activities. They are:

- **Procurement (purchasing)**: This is what the organization does to get the resources it needs to operate. This includes finding vendors and negotiating best prices.
- **Human resource management**: This is how well a company recruits, hires, trains, motivates, rewards, and retains its workers. People are a significant source of value, so businesses can create a clear advantage with good HR practices.
- **Technological development**: These activities relate to managing and processing information, as well as protecting a company's knowledge base. Minimizing information technology costs, staying current with technological advances, and maintaining technical excellence are sources of value creation.
- **Infrastructure**: These are a company's support systems, and the functions that allow it to maintain daily operations. Accounting, legal, administrative, and general management are examples of necessary infrastructure that businesses can use to their advantage.

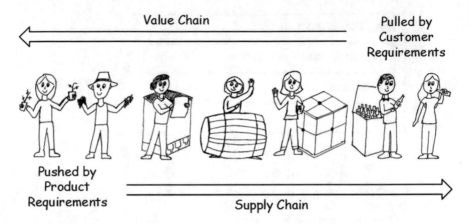

Fig. 4.6 Value chain vs. supply chain

As a rule of thumb, primary activities are value added activities, while support activities should be at least non-value added activities, but not waste.

A supply chain and a value chain are complementary views of an extended enterprise with integrated business processes enabling the flows of products and services in one direction, and of value as represented by demand and cash flow in the other (Fig. 4.6). Both chains overlay the same network of companies. Both are made up of companies that interact to provide goods and services. When we talk about supply chains, however, we usually talk about a downstream flow of goods and supplies from the source to the customer. Value flows the other way. The customer is the source of value and value flows from the customer in the form of demand to the supplier. That flow of demand, sometimes referred to as a "demand chain," is manifested in the flows of orders and cash that parallel the flow of value and flow in the opposite direction to the flow of supply.

Thus, the primary difference between a supply chain and a value chain is a fundamental shift in focus from the supply base to the customer. Supply chains focus upstream on integrating supplier and producer processes, improving efficiency and reducing waste, while value chains focus downstream on creating value in the eyes of the customer. All chains in the value chain must support the final value delivery. This distinction is often lost in the language used in the business and research literature.

 4.6 A Practical View

According to Lean Philosophy, value can only be seen by the eye of the beholder. Our definition of value encompasses all the dimensions that the stakeholders might consider important (valuate): it goes beyond the product functionalities and/or the service functions, including aspects like time (to market), cost, risk acceptance etc. Therefore, no real value can be pushed but only pulled through the value chain.

Even though it is easy to understand the previous statement, it is hard to put into practice:

- We often have scarce time to invest in understanding what is expected from our internal and external stakeholders. We tend to use preconceived ideas and wishful thinking to define the value.
- Even when we have the time and resources, understanding the stakeholders is not an easy task, particularly because their vision of reality might be quite different from ours. Standing in somebody else's shoes is not trivial.
- The real value is subtle and normally not verbalized, thus it cannot be heard, but only felt.

When we fail to deliver the expected value we either loose the customer or are inserted into rework cycles. We like to joke that: "**We never have the time to do it right, but we always have the time to do it twice!**"

Very often, companies fail to listen to all the relevant external and internal stakeholders in order to understand the complete product development program value.

External stakeholders are the ones the pull value from the product development program's final results (the product and/or services). They can be encountered when we consider the "Product/Process Follow-up" and the "Product Discontinuation" process groups from the product development process (Fig. 4.4). Some examples of external stakeholders are:

- **User**: the end user/consumer of the product and/or service developed.
- **Customer** (sometimes known as a client, buyer, or purchaser): the recipient of the good, service, or product resulting from the development, which is obtained from a seller, vendor, or supplier for a monetary or other valuable consideration. We do not consider here the intermediate customers or trade customers (more informally: "the trade") who are dealers that purchases goods for re-sale, but the ultimate customers.
- **Shareholder/Sponsor**: who pays all or part of the development cost and expects something (value) as the return from this investment.
- **Dealer**: a person or firm engaged in commercial purchase and sale the good, service, or product resulting from the development.
- **Distribution logistics**: The focal point of distribution logistics is the shipment of goods from the manufacturer to the consumer. Logistics comprises all activities related to the provision of finished products and merchandise to a customer.
- **Training network**: a person or firm engaged in training the users so they can use correctly and perceive the complete value from the good, service, or product resulting from the development.
- **Maintenance and repair services**: a person or firm engaged in providing maintenance services in order to keep/return the good, service, or product resulting from the development to its ideal value delivery condition.

- **Recycler (eco-friendly)**: a person or firm engaged in recycling the good, or product resulting from the development.
- **Regulatory agencies**: A regulatory agency (also regulatory authority, regulatory body or regulator) is a public authority or government agency responsible for exercising autonomous authority over some area of human activity in a regulatory or supervisory capacity. An independent regulatory agency is a regulatory agency that is independent from other branches or arms of the government.
- **Others**: any other external stakeholder relevant for your particular development project.

Internal stakeholders, by the other hand, are the ones who pull value from the product development program intermediate results. They can be encountered when we consider the "Design & Development" and the "Production/Ramp-up" process groups from the product development process (Fig. 4.4). Some examples of internal stakeholders are:

- **Shareholder/Sponsor**: who pays all or part of the development cost and expects something (value) as the return from this investment. As well as expecting value from the product itself, they might also pull value from the product development value stream.
- **Suppliers**: the suppliers of raw material, components, or any material necessary to produce de product or perform the service resulting from the development.
- **Design and development team**: the functional areas inside the development organization which are engaged in the product's design and development activities.
- **Production**: the production area inside or outside the organization, which will be engaged in the product's production.
- **Development partners**: a person or firm that partners on the product's design and development activities.
- **Quality**: the quality area inside the organization, which will be engaged in the quality management activities related to the product and the PDP.
- **Tests**: the test area inside the organization, which will be engaged in testing the product while it evolves into its final version though the PDP.
- **Distribution logistics**: the area inside the organization which is engaged in the activities related to the provision of finished products and merchandise to a customer.
- **Recycler (scrap)**: a person or firm engaged in recycling the scrap and/or c-products resulting from the development.
- **Regulatory agencies**: the same as in external stakeholders, considering any regulations related to the product development activities themselves.
- **Others**: any other internal stakeholder relevant for your particular development project.

While failing to consider external stakeholders means that the product will face issues after launch, failing to consider internal stakeholders means the development process will not flow as smoothly as it could be.

References

1. American Heritage® Dictionary of the English Language (2015) online. https://www.ahdictionary.com/
2. Fleming QW, Koppelman JM (2005) Earned value project management, 3rd edn. Project Management Institute: Newton Square
3. Project Management Institute, PMI (2013) A Guide to Project Management Body of Knowledge (PMBOK® Guide), 5th edn. Project Management Institute, Newton Square
4. Bonnal P, De Jonghe J, Ferguson J (2006) March) A deliverable-oriented EVM system suited to a large-scale project. Project Manage J 37(1):67–80
5. Csillag JM (1985) Análise do valor. São Paulo, Editora Atlas
6. Ohno T (1998) Toyota production system. Productivity Press, New York
7. Womack JP, Jones DT (2003) Lean thinking. Free Press, New York
8. Morgan JM, Liker JK (2006) The Toyota product development system. Productivity Press, New York
9. Mascitelli R (2002) Building a project-driven enterprise. Technology Perspectives, Northridge
10. Murman et al (2002) Lean enterprise value: insights from MIT's lean aerospace initiative. Polgrave, New York
11. Loureiro G (1999) A systems engineering and concurrent engineering framework for the integrated development of complex products. Ph.D. Thesis, Department of Manufacturing Engineering, Loughborough University: Loughborough, UK
12. Porter ME (1985) Competitive advantage: creating and sustaining superior performance. The Free Press, New York

Chapter 5
Waste in Product Development

The opposite from value is waste. Waste absorbs resources, increase cost and create no value. Therefore the result from a wasteful activity is something that no one wants to pay for. Even though there is no Toyota definition of waste outside of the production process, the notion of waste helps to understand their development system. In fact, there are several adaptations from the original seven wastes, most of them varying on the descriptions and including additional wastes to the set. This chapter compares some of these proposed sets and presents the 10-waste set used in this book. We also discuss how each of these 10 wastes relate to the Product Development System elements and why we consider creating a waste-free and "waste proof" process an unfeasible task. Even though it is unlikely that the wastes will be completely eliminated, they can be considerably reduced.

5.1 Introduction

In "lean terms" low performance is the consequence of waste.

> Waste refers to all elements of a process that only increase cost without adding value, or any human activity that absorbs resources but creates no value. [1, 2].

The seven original manufacturing wastes are[1]:

1. **Waste of overproduction**: The result of producing items for which there are no orders generates such wastes as overstaffing and storage and transportation costs because of excess inventory.

---

[1]Waste list from Ohno [1, p 19], and definitions from Liker [3, p 29–30].

© Springer International Publishing AG 2017
M.V.P. Pessôa and L.G. Trabasso, *The Lean Product Design and Development Journey*, DOI 10.1007/978-3-319-46792-4_5

2. **Waste of time in hand (waiting)**: This occurs when workers merely serve to watch an automated machine or have to stand around waiting for the next processing step, tool, supply, part, etc., or just plain have no work because of stock outs, lot processing delays, equipment downtime, and capacity bottlenecks.
3. **Waste in transportation**: This is the result of carrying work in-process (WIP) long distances, creating inefficient transport, or moving materials, parts, or finished goods into or out of storage or between processes.
4. **Waste of processing itself**: The processing is inefficient due to poor tool and product design; it takes unneeded steps to process the parts.
5. **Waste of managing stock (inventory)**: This occurs when excess raw material, WIP, or finished goods cause longer lead times, obsolescence, damaged goods, transportation and storage costs, and delay. Also, extra inventory hides problems such as production imbalances, late deliveries from suppliers, defects, equipment downtime, and long setup times.
6. **Waste of movement**: This includes any wasted motion employees have to perform during the course of their work, such as looking for, reaching for, or stacking parts, tools, etc. Also, walking is waste.
7. **Waste of making defective deliverables**: Repair or rework, scrap, replacement production, and inspection mean wasteful handling, time, and effort.

Even though there is no Toyota definition of waste in product development, the notion of waste helps to understand their development system [4]. Fortunately, the original waste definition can be generalized to other domains such as product development, order taking, and the office [1]. In fact, there are several adaptations from the original seven wastes, most of them varying on the descriptions and including additional wastes to the set. Instead of being a deviation from the original thinking, these changes much more reflect the different expected use of the proposed sets.

Table 5.1 presents some waste sets from the literature plus the one used in this book. The scope (manufacturing, general, information, or product development— PD) of each set is highlighted. Each line shows the more compatible definitions across authors. Note that in several cases more than one of an author's types has been captured in one of the types defined in this work. For example, Liker's [3] "unused employee creativity" is captured in Inventory.

Wastes themselves affect the Product Development System's elements creating an intricate net. Understanding this net requires knowing how each of the system's elements might have its performance degraded. To achieve this objective, we decided to use a waste set that is a merging of those mentioned previously. Instead of trying to translate the waste definitions from the production system to the product development system, this book assumes that the waste causes the deterioration of the PDS. A set of 10 waste drivers was considered in order to better link the drivers to the PDS elements (Fig. 5.1).

The choice and organization of the waste drivers do not greatly differ from what is presented in the literature. Whenever possible, the original waste nomenclature was maintained, in order to avoid misinterpretations and

Table 5.1 Comparison of waste sets

Ohno [1]	Womack and Jones [2]	Liker [3]	Bauch [5]	Kato [6]	Morgan and Liker [7]	McManus [8]	Ward [4]	This work
manufacturing	general	general	PD	PD	PD	PD	Information	PD
Overproduction	Overproduction	Overproduction	Overproduction/ Unsynchronized processes	Overproduction of Information (Duplication)	Overproducing	Overproduction		Overproduction
Time on hand (waiting)	Waiting	Waiting	Waiting	Waiting of people	Waiting	Waiting		Waiting
Transportation	Transport	Unnecessary transport or conveyance	Transport/ Handoffs	Transportation of Information	Conveyance	Transportation		Transportation
Processing itself	Processing	Overprocessing or incorrect processing	Overprocessing	Overprocessing	Processing	Overprocessing		Overprocessing
Stock on hand (inventory)	Inventories	Excess inventory	Inventory		Inventory	Inventory		Inventory
Movement	Movement	Unnecessary movement	Movement	Motion of People	Motion	Unnecessary movement		Motion
Making defective products	Defects	Defects		Rework	Correction	Defective products		Correction
	"Wrong design"		Defects	Defective information	Defects			Defects
		Unused employee creativity						Captured in Inventory
			Re-invention	Re-invention				In Overprocessing

(continued)

Table 5.1 (continued)

Ohno [1]	Womack and Jones [2]	Liker [3]	Bauch [5]	Kato [6]	Morgan and Liker [7]	McManus [8]	Ward [4]	This work
			Lack of system discipline					In Overproduction and Overprocessing
			Limited IT resources					In Overproduction and Motion
				Hand-offs			Hand-offs	In Transportation
							Scatter	In Happenings
							Wishful thinking	Wishful thin king
								Happenings

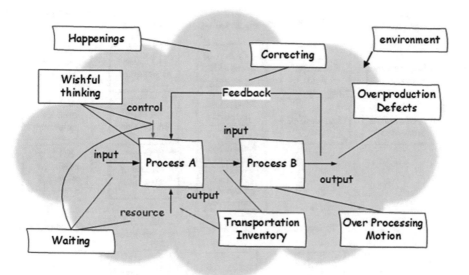

Fig. 5.1 PDS elements and low performance drivers

misunderstandings. The most relevant contribution on the set is the inclusion of "Happenings," as a waste type rooted in the external environment. (as identified by Gershenfeld and Rebentisch [9]).

Each of the 10 waste types have subtypes (Fig. 5.2) that better define their scope. Indeed, the root causes of unscheduled waste (caused by variations from the planned) differ from the root causes of scheduled waste (normally a result of bad planning or the consequence of resource allocation restrictions). The 10 waste types can be summarized as:

1. **Overproduction**: Producing process outputs at a higher rate or earlier than the next process can use them is overproduction; its subtypes are unnecessary processes and unsynchronized processes.
2. **Waiting**: This refers to the part of processing time when the creation of value remains static, hence the value stream is considered as 'non-flowing' due to the lack of necessary inputs, resources or controls.
3. **Transportation**: This includes the loading, transporting, and unloading of outputs/inputs (information or material) and resources from place to place without adding value during the process.
4. **Over processing**: Completing unnecessary work during a process is considered over processing.
5. **Inventory**: This includes raw, in-process or finished buildup of information, knowledge, or material such as prototypes that are not being used.
6. **Motion**: This refers to any unnecessary movement of people or activity during non-transformation task execution in a process.

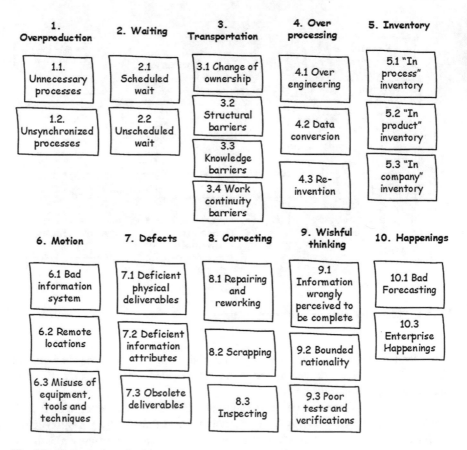

Fig. 5.2 Waste types and subtypes

7. **Defects**: Defects are the creation of defective outputs as a result of the development process.
8. **Correcting**: This is the result of redoing or scrapping, due to feedback. Correcting subtypes are repairing/reworking, scrapping, and inspecting to find problems.
9. **Wishful thinking**: This means making decisions (mental activity) without the needed inputs (data) or operating according to incorrect controls.
10. **Happenings**: This includes all reactions to unexpected happenings in the environment.

5.2 Overproduction

Overproduction is perceived only from the perspective of the next process. A PDS incurs overproduction whenever the previous processes deliver superfluous inputs to processes (unnecessary processes), or release inputs in a higher rate or before the next processes can handle (unsynchronized processes).

5.2.1 Unnecessary Processes

Unnecessary processes include any forced process output that is not necessary (which are different from defective, but needed, deliverables) and have to be sorted from useful work. Not needed deliverables might be the result of doing duplicate work or simply the creation of unnecessary deliverables.

Duplicate work happens when the company or the development team structure has redundant functions, the division of labor is unclear, there is insufficient communication and coordination, or even due to the company/team inability to prompt adjustment of the division of labor. During structural or process changes, people have to relearn the network of communications, understand how their components fit into the system, and recreate the real process of working together [4]. Change resistance and the lack of specific training on the new way of doing also play a role on the delay to the adjustment of the division of labor. Finally, there is the system inertia and the time for the news to spread throughout the organization.

Unnecessary processes might also be a consequence of a bad standard process, a bad contracted Statement of Work (SOW), or by a bad development plan that defines deliverables which are not needed. The team members themselves may also execute unnecessary processes (which differ from over engineering, where the delivered are still needed but are "gold plated") by their own sake on a hidden agenda basis or as a protection against uncertainty.

5.2.2 Unsynchronized Processes

Unsynchronized processes mean that the delivered process outputs will not be promptly used because of lack of capacity (excessive) or because other elements needed in order to proceed are unavailable (inopportune).

Unsynchronized processes are either the consequence of a poorly planned schedule or the result from issues during the development execution. A schedule may be originally unsynchronized due to a non-optimized standard process, the lack of the needed resources to define a smooth work flow, or simply by bad planning. Information batch processing can also overload the next process once idle times are followed by waves of work to be executed. Furthermore, execution rarely occurs as planned. Product development is intrinsically uncertain, not identified risks or changes might occur and its chain effect might disrupt the originally synchronized plan.

5.3 Waiting

Waiting means that the system is idle ('non-flowing') expecting some needed: (1) authorization to perform the work; (2) input to be processed; or (3) resource to be used during execution. Waiting for authorization means that, although the process has all the necessary inputs and resources, it is not authorized to start processing. Authorization might be a pre-defined moment or some control input to trigger the process. Waiting for input or resource means that the previous processes did not deliver its work on time or that some other process is keeping the necessary resources more than initially expected. Wait can be "scheduled" or "unscheduled."

5.3.1 Scheduled Waiting

In the case of scheduled waiting, people, information, or resources are planned to stay idle during some time. Wait times are input during planning as a consequence of:

- **Excessive buffer time added by the planner**: More than necessary reserve time is included between activities due to perceived risks, historical delays, or imposed by the standard process and guidelines (i.e. buffer must be equal to X % of the critical path).
- **Lack of resources**: There are no available resources to perform parallel independent tasks.
- **Uncertainty of resource availability**: Worst case scenarios are used to reduce risks, tough turning the schedule longer.
- **Existence of dependency between tasks**: The net interdependence between tasks does not allow a wait free plan.
- **Long or unpredictable internal/external lead times**: Internal and/or external (suppliers, regulatory agencies, etc.) lead times are either long or unpredictable, imposing wait to possible parallel tasks, or the use of large buffers.

5.3.2 Unscheduled Waiting

Unscheduled waiting is the unexpected wait time that occurs during the development due to:

- **People neglect the schedule**: The scheduled is either neglected or not enforced; the student syndrome (leaving things to be done only in the last moment) is an example of risky behavior [10].
- **Changes**: Changes causing the duration of the activities differ from the originally planned activities.
- **Planned schedule is too tight**: The schedule is unrealistic and fated to delays; while excessive buffer means planning wait, too tight schedules results in unscheduled wait. Too-tight schedules lead to a complete lack of directions during execution since they are quickly thrown away.
- **Resource performance below the expected**: Resource performance estimation may have been pure wishful thinking. Having no replacement or lack of maintenance leads to fatigue of people, machines, or communication channels. Finally, the resource (person or equipment) assigned may not fit the job, causing delays.

- **Cascade waste**: Other wastes, like overproduction, transportation, over process-ing, motion, happenings, etc., have the power to disturb the flow, causing wait.

5.4 Transportation

Transportation includes the loading, transporting, and unloading of outputs/inputs (information or material) and resources from place to place without adding value during the process. Transportation happens whenever information or materials change ownership or have to overcome structural barriers. Transportation also occurs when information has to be "loaded and unloaded in a person" due to knowledge barriers (need to learn) or to continuity barriers (interruptions, multitasking, etc.).

5.4.1 Change of Ownership

Change of ownership has two main causes: unclear responsibility or authority, and hand-offs.

By not having a clear assignment of responsibility or authority, people keep sending and receiving pieces of work as a way for not be blamed for failures and mistakes. Due to lack of authority, people ask permission to continue his work or release deliverables, regardless what is defined on the standard process (if there is one). Hand-offs, however, means that the ownership is being changed to follow a process or plan that detaches knowledge, responsibility, feedback, and action [4].

5.4.2 Structural Barriers

Structural barriers result from the bad distribution of people or physical resources, or by an inexistent or non-reliable communication channel. A bad physical distri-bution imposes transportation in order to allow the process to be executed by the necessary resources. Issues related to the communication system may require the use of alternate ways, such as manual handling, face-to-face interactions, etc., or the existence of non-optimal workflows.

5.4.3 Knowledge Barriers

There is a knowledge barrier whenever a person does not have the necessary knowledge to perform a task, needing to acquire it from the basics or from the practice. Knowledge barriers require transportation of people and/or resources to perform the needed training.

5.4.4 Continuity Barriers

Work continuity barriers are caused by interruptions that require the person's train of thought to change direction. They require unloading the current information, loading the new data, processing, unloading the no longer useful information, and reloading to the original state. The need of unscheduled input (solving doubts and problems inside or outside the project) may trigger an interruption and the consequent stop and go effect, where the engineer has to reorient himself to a certain task and it is like a setup for a machine [11]. Multitasking and task switching inside the project, between projects, or between functional and project activities have the same effect of requiring a change of a "mind setup."

5.5 Over Processing

While overproduction is related to the output of the process, over processing includes completing unnecessary work during a process. Over processing can be divided into: over engineering (beyond what the specifications require), data conversion (converting data between information systems or between people), and the re-invention of anything that could be readily reused or adapted.

5.5.1 Over Engineering

When specifying too much detail, the designer wastes time, and also can set unnecessarily rigid tolerances that constrains the development. Even though normally associated to a perfectionist personality (gold plating), it can be a consequence of the lack of knowledge of the expected level of detail. The needed level of detail may not be known due to: the individual designer's poor understanding of downstream tasks (what comes next and its needs), the lack of constructive advice from experienced designers, and the downstream designers not helping upstream peers release the right level of detail on deliverables.

The over engineering of a deliverable can result from the lack of confidence or even from the perfectionist personality of a designer. A badly defined standard process or task execution guidelines may also require excessive processing.

5.5.2 Data Conversion

Data conversion includes both converting to different measurement systems and translating to other languages. The former may input errors due to rounding, and the latter can change the meaning of the information being transmitted.

Nonstandard data format use among designers, and incompatible information systems and tools or version-ups increase the need of conversions.

5.5.3 Re-Invention

Re-invention of processes, solutions, methods, and products which already exist or rather would only require some modifications to make them fit for the use at hand may be a consequence of sheer lack of knowledge of the legacy existence, or not realizing the benefits of starting from the same level of knowledge of past developments and choosing to do everything again [11]. The lack of knowledge can be a result of: poor expertise sharing, a bad knowledge management system, security issues that may prevent expertise sharing, the designer's unwillingness to share their expertise, or even the abandonment of past experiences and lessons learned. If the next process in the value chain receives both the original and the re-invented as inputs the re-invention is also a cause of overproduction.

5.6 Inventory

Inventory appears between processes (outputs), in the corporate environment, and even inside the outputs. Inventories can be found in the company as equipment or data storage, between processes as work in-process, or inside the deliverables as excessive information, components, or design options. While the company spends money and resources to keep this material, they also incur on the risk of this work.

5.6.1 In-Process Inventory

In-process inventory means material and information held between or within processes' activities. In-process inventory happens when the processes are unable to promptly handle all the received information or materials and is the result of: high system variability, exceeding capacity utilization, or batch sizes.

5.6.2 In-Product Inventory

In-product inventory is the result of unnecessary features, options (i.e. capability to add sections to an aircraft hull) or parts included to systems and subsystems. Features are added even though nobody is interested in paying for them. They might be added for the designer's sake, or be the result of unclear/shifting goals or insufficient pervasion of goal information (insufficient time to read/examine release information, spatial/structural barrier, etc.). In-product inventory may imply more failure modes, more needed room in the product, higher weight, and additional time and costs to design, test, and produce.

5.6.3 In-Company Inventory

In-company inventory consists of sub-utilized or unnecessary equipment and prototypes, and excessive data storage. This kind of inventory consumes resources and space without adding value. Worse, excessive data storage means more time to find the useful information among pure trash.

5.7 Motion

Motion differs from (1) transportation in the sense that the former considers only the movement of the performer while the latter focus on the transportation of materials and information; and (2) over processing because it considers movements that do not transform inputs into deliverables. Motion can be typified as unnecessary human motion due to bad information systems, remote locations, and not optimized use of equipment, tools, and techniques by not understanding them (too complex, lack of training).

5.7.1 Bad Information System

A badly designed information system may contribute to motion by not allowing the needed and available information to be directly accessible by the user, or by requiring time consuming searches (hunt) to be found. People have to either leave their workplace to make a physical search, or directly ask people, or search through the project directory structure on a server [5].

5.7.2 Remote Locations

The information owner or its storage place is not directly accessible from the working environment. The local distance of departments and facilities often has negative impacts on the project team work: (1) there is the loss of time required to move to and from the remote location; (2) the remoteness indirectly acts as kind of a barrier and discourages people from making that trip; and (3) remoteness inhibits the formation of productive teams [5].

5.7.3 Misuse of Equipment, Tools, and Techniques

Whenever they not mastered, equipment, tools, and techniques are misused. Lack of training (theoretical or practical) or sheer complexity are reasons mastery can be difficult. "Complex" in this case means having non-intuitive operation procedures, badly designed interfaces and interface navigation, or difficult to understand instructions. Complex equipment, tools, and techniques take more time to be mastered, require more steps to be used, and are more easily misused than a simpler alternative.

5.8 Defects

By *defects* we mean the creation of defective outputs from the development process. Defects are perceived as deficient physical deliverables, deficient information, or information that becomes obsolete while in process.

5.8.1 Deficient Physical Deliverables

Physical deliverables include not only the final product, but also the parts and subsystem that are created through the development project. A physical deliverable may be defective for several reasons such as:

- The legacy of defects from previous and reused versions, where some issues remain dormant and only awaken when new features are added or previous features are stressed;
- "Card castle" design that means that the current design is not robust. This may be the consequence of "deficient information attributes", too many lapses, a complex architecture, unstable technology, etc.
- Poor/inefficient tools that people have at hand or are required to use.

5.8.2 Deficient Information Attributes

Strong et al. [12] suggested four categories with a total of 15 different attributes describing information quality IQ:

- **Intrinsic IQ**: Accuracy, objectivity, believability, reputation
- **Accessibility IQ**: Accessibility, security
- **Contextual IQ**: Relevancy, value-added, timeliness, completeness, amount of information

- **Representational IQ**: Interpretability, ease of understanding, concise representation, consistent representation

Deficiencies in one or more of these attributes do not necessarily mean that the information at hand lapses immediately and becomes useless; some deficiencies may be even compensated by the designer's knowledge and experience. Unperceived deficiencies, though, may induce wrong decisions and wishful thinking.

5.8.3 Obsolete Deliverables

Obsolete deliverables, although not deficient when released, become obsolete while waiting to actually be used. Long lead times and excessive waits might be reasons to obsolescence.

5.9 Correcting

Correcting is the redoing or scrapping due to feedback. Correcting subtypes are repairing/reworking, scrapping, and inspecting to find problems.

5.9.1 Repairing and Reworking

Repairing and reworking aim to correct or optimize what has already been done. They are mainly driven by: optimization process or "refactoring" of work, poor/incomplete information from previous phases (deliverables that do not match perfectly, unnecessarily tight tolerances that need to be revised and legacy parts or information that do not fit properly), patch work to fix workaround solutions, detected defects, and changes. Repairing and reworking, although actually adding value, are considered as waste. They are wasteful because ideally they should not occur if things were done right at the first time.

5.9.2 Scrapping

If the defective deliverables cannot be repaired they have to be done again, with loss of material and/or time.

5.9.3 Inspecting

Inspecting includes resources that are used to find defects instead of effectively adding value. Unreliable processes are the main reason for inspecting. The greater the development uncertainty and complexity, the higher the necessary inspecting to keep track of the progress.

5.10 Wishful Thinking

Wishful thinking means making decisions (mental activity) without the needed input (data), or operating according to incorrect controls. Subtypes of wishful thinking include: decisions made using information wrongly perceived to be complete, decisions biased by bounded rationality, or the execution of poor tests and verifications that do not guarantee the value delivery.

5.10.1 Information Wrongly Perceived as Complete

People believe that incomplete information is complete when:

- Complex products, processes, organizations, markets or business prevent a complete picture of the reality. The complexity also imposes a higher number of variables to be taken into account, hardening inference making and decision making.
- Unclear and shifting goals make it difficult to picture the reality.
- The information is not available on the system in a timely manner. The information may not be available at all or its delivery may be unsynchronized to the demand (late delivery).
- Lack of time imposes the decision making based on the available information so far, regardless of quality and completeness. The time pressure may be internally imposed by the schedule, may be a consequence of previous delays, or can result from organizational, market, and business changes.

5.10.2 Bounded Rationality

Even though the information is available and there is no time pressure, bounded rationality prevents the clear decision making. The rationality may be bounded by: personal reasons such as lack of knowledge, lack of discipline, prejudice, pride, etc.; organizational culture that prevents doing things differently from the way things are traditionally done; and impositions from higher management or bad contracts.

5.10.3 Poor Tests and Verifications

Wrong testing or the bad execution of well-designed tests gives misguiding results. Poor tests and verifications are done when: testing only to specifications and not to failure, by not considering all the failure modes; excessive optimism, wishful thinking, risk taking strategy, and bounded rationality may cause the alarms to be suppressed; and defects from previous versions, reused modules, etc. that end up not being covered by the test set.

5.11 Happenings

Happenings include all reactions to unexpected happenings in the environment. Happenings result from failing to forecast the changes in the market and in the business, or from changes of the internal environment (structure, rules, etc.). Reacting to unexpected happenings can trigger a wave of changes that hugely affect the development project performance.

5.11.1 Bad Forecasting

Bad forecasting means that the company will need to react to unexpected happenings on the business and on the market because it was unable to foresee the changes. The understanding of the market is achieved by market research and market intelligence. Business forecasting can be made, for instance, through scenario analysis considering political factors, economical factors, social factors, labor factors, other factors from the business environment.

5.11.2 Enterprise Happenings

Enterprise happenings are one great source of discontinuity in the development. The development team must react to happenings due to:

- The change of the enterprise priorities, shifting the project and either imposing more time pressure or reducing available resources.
- Emergencies that are a special case of change of priorities where task forces are formed to "put out fires" in other parts of the company or other projects. Quite frequently, when the task forces are disbanded the "fire fighters" discover that their own projects are now "on fire."

- Reorganizing that breaks all the already formed communication channels and triggers chain reactions that take precious time away from the development system to break the inertia and become stable again.
- The overloading of the PDS by the addition of more projects than it is capable to deal with.
- Adding formal structure on the conventional management assumption that the creation of order is done through organizational structure (procedure manuals, organization charts, and directives), instead of recognizing that order emerges from interaction among the people, which takes time [4].

 ## 5.12 A Practical View

Once the low PDP's performance is the consequence of waste, the waste occurrence erodes its two pillars of "do the right product" and "do the product through the right process."

All the wastes, in a way or another, impact the development (time, cost, productivity, and capability) indicators (Table 5.2). Since they absorb resources in a way that do not add value, the development takes longer and is more expansive than it could be. Development productivity and the whole development organization capability also suffer from this inefficiency. This is the reason why waste reduction and elimination is paramount to achieving a high performance product development system [13].

Wastes themselves affect each other, creating an intricate net, and understanding this net requires knowing how each of the wastes is influenced by (passive role) or influences (active role) the others in the set.

Table 5.2 Wastes versus PDP indicators

	Product quality	Product cost	Product business case	Development time	Development cost	Development productivity	Development capability
1. Overproduction				⇨	⇨	⇨	⇨
2. Waiting			⇨	⇨	⇨	⇨	⇨
3. Transportation				⇨	⇨	⇨	⇨
4. Over processing				⇨	⇨	⇨	⇨
5. Inventory				⇨	⇨	⇨	⇨
6. Motion				⇨	⇨	⇨	⇨
7. Defects	⇨	⇨	⇨	⇨	⇨	⇨	⇨
8. Correcting				⇨	⇨	⇨	⇨
9. Wishful thinking	⇨	⇨	⇨	⇨	⇨	⇨	⇨
10. Happenings		⇨	⇨	⇨	⇨	⇨	⇨

Creating a waste-free and "waste proof" process is an unfeasible task [14]. This is the consequence of the "Happenings" waste category which has no causes within the waste set, since they represent the environment's influence on the system. Their existence and relevance document the difficulty of having a waste-free PDS. Market and business risks are hard to predict and are the causes of "bad forecasting." Even though it is unlikely that these wastes will be completely eliminated, they can be reduced. This fact also enforces the need of continuous improvement which is the next chapter's subject.

References

1. Ohno T (1998) Toyota production system. Productivity Press, New York
2. Womack JP, Jones DT (2003) Lean Thinking. Free Press, New York
3. Liker JK (2004) The Toyota way: 14 management principles from the world's greatest manufacturer. McGraw-Hill, New York
4. Ward A (2007) Lean product and process development. The Lean enterprise Institute, Cambridge
5. Bauch C (2004) Lean product development: making waste transparent. Thesis (Diploma) Massachusetts Institute of Technology, Cambridge
6. Kato J (2005) Development of a process for continuous creation of lean value in product development organizations. Thesis (Master) Massachusetts Institute of Technology, Cambridge
7. Morgan JM, Liker JK (2006) The Toyota product development system. Productivity Press, New York
8. McManus H (2004) Product development value stream mapping (PDVSM) manual. Beta draft, MIT Lean Aerospace Initiative, Cambridge
9. Gershenfeld J, Rebentisch E (2004) The impact of instability on complex social and technical systems. Proceedings from MIT engineering systems division external symposium
10. Goldratt EM (1997) Critical chain. Hampshire, UK, Gower Pub Co
11. Morgan JM (2002) High performance product development: a systems approach to a lean product development process. (Thesis)—Industrial and Operations Engineering: University of Michigan
12. Strong D, Lee Y, Wang R (1997) 10 potholes in the road to information quality. Computer 30(8):38–46
13. Pessôa MVP (2008) Weaving the waste net: a model to the product development system low performance drivers and its causes. Lean Aerospace Initiative Report WP08-01, MIT: Cambridge, MA
14. Pessôa MVP, Seering W, Rebentisch E, Bauch C (2009) Understanding the waste net: a method for waste elimination prioritization in product development. In: Chou S et al (Org) Global perspective for competitive enterprise, economy and ecology. London: Springer, pp 233–242

Chapter 6
Continuous Improvement

This chapter discusses how the continuous improvement acts like the motor that keeps the lean wheel running. As we stated in the previous chapters, while lean product development wheel goes toward its goal (the value to be delivered) all types of issues come on the way and might cause waste. Continuous improvement prevents a process from getting stuck or even slipping back. Any process is measured by its performance (process metrics) and its results (deliverables/products metrics). The PDP differs from repetitive processes/tasks once it aims to produce a new product each time it is executed; also, the PDP has to be shaped according to the particular product it aims to produce. Therefore the PDP itself can be considered a continuous improvement process, which gives an even greater importance to the aspects presented in this chapter. Here we discuss the continuous process and product improvement.

6.1 Introduction

Ohno [2] defines three types of work:

1. **Value-added work**: The processing that adds value in the sense the customer perceives it.
2. **Pure waste**: The processing that only increases cost without adding or supporting any value.
3. **Non-value added work**: Things that, even though they do not create value themselves, have to be done under the present work conditions to support the value-added work.

By definition, *improve* means "raise to a more desirable or more excellent quality or condition." Continuous improvement is the ability to move from the actual state to a new desired, recognizing the path ahead as unclear and unpredictable, requiring sensitiveness and responsiveness to actual conditions on the ground [1].

© Springer International Publishing AG 2017
M.V.P. Pessôa and L.G. Trabasso, *The Lean Product Design
and Development Journey*, DOI 10.1007/978-3-319-46792-4_6

The continuous improvement acts on eliminating waste and on changing the work conditions in a way to reduce the non-value added work to the minimum.

From the previous definition, three aspects can be identified:

- New desired state (target condition)
- Unclear territory
- Sensitive and responsive

The culture and capability for continuous, incremental evolution and improvement represents perhaps the best assurance of durable competitive advantage and company survival.

Sections 6.2 and 6.3 are adaptations from Rother [1], which we highly recommend as a comprehensive reference of how to perform the continuous improvement practice.

6.2 Continuous Improvement

Briefly put, the improvement goes like this (Fig. 6.1) [1]: (1) In consideration of a vision, which defines a long term final target condition, and (2) with a firsthand grasp of the current condition, (3) a next intermediate target condition on the way to the vision is defined. When we then (4) strive to move step by step (sensitive to the unclear path ahead) toward that target condition, we encounter and face (responsiveness) the obstacles that define what we need to work on, and from which we learn.

The responsiveness is the essence of Jidoka (see Chap. 3), where deviations from the path to the target condition (finding an obstacle) trigger a continuous improvement cycle.

6.2.1 Target Condition

First and foremost, improvement is only taken when we go towards a predefined target condition or vision. A target condition describes a desired future state: the

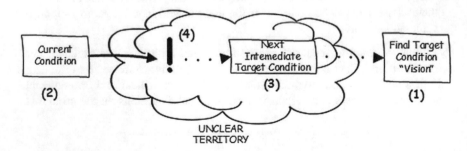

Fig. 6.1 Moving through unclear territory

situation we want to have in place at a specific moment in the future time. The target condition acts like a beacon and all improvement efforts are aligned towards it. Therefore, changing is only considered improvement when it means climbing a step closer to this vision.

The target condition must be on the medium-to-long term and be challenging, so it can give direction to all improvement efforts. Failing to do that will reduce the likelihood of the alignment of concurrent improvements efforts as well as hinder the sequence of actions will lead to a clear goal. Shorter term and easier to achieve target conditions might support achieving the long term one since they build a series of steps that facilitate the climbing towards it (Fig. 6.2):

- Final target condition = Vision = Long/Medium term
- Intermediate target condition = Step = Short term

Once a target condition is defined, the condition itself is neither optional nor easily changeable. It stands. How to achieve that condition *is* optional (people, place, culture, etc. may change the "how") [1]. The sequence of intermediate target conditions guarantees the continuous motion of the continuous improvement approach.

Having a target condition is paramount for effective process improvement; the company should not start trying to improve or move forward before a target condition has been defined.

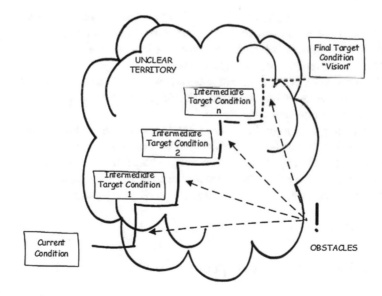

Fig. 6.2 Climbing the stairway

6.2.2 Unclear Territory

The path from the current to the target condition is always unclear territory. The target condition will function like a beacon, but we will find obstacles on the way (Fig. 6.1). By working through the obstacles, while heading towards the vision, real improvement is achieved. Indeed, any time you establish a target condition and try to move toward it, you will discover problems and obstacles. Then you have two choices:

1. Avoid the obstacle(s) and deviate from the vision (or change the vision—normally due to a weak vision).
2. Work through the obstacle(s) by understanding and eliminating its causes (this is the lean way choice).

If the way ahead is assumed clear, we tend to do wishful thinking and blindly carry out a preconceived implementation plan rather than being sensitive to learning from and dealing adequately with what arises along the way. As a result, we do not reach a desired destination at all, despite our best efforts.

6.2.3 Responsiveness

Responsiveness is the ability to adjust quickly to suddenly changed external conditions, or abnormal conditions, and to resume stable operation without undue delay. Considering that our stable operation is going towards the target condition, we should be responsive to any obstacle in its way.

Being responsive is not implementing preconceived steps or solutions, which may or may not work as intended, but understanding the logic and method of how to proceed through unclear territory.

Understanding this logic has more importance than the solutions themselves: the company's ability to be competitive and survive lies not so much in solutions themselves but in the capability of its people to understand a situation and responsively develop solutions [1]. The organization must have the capability of keeping forward movement, improving, adapting, and satisfying dynamic customer pulled value.

6.3 Continuous Process Improvement

We divide the continuous improvement into continuous process improvement and continuous product improvement. This section discusses the traditional improvement approaches and how they compare to the lean continuous improvement.

6.3.1 Traditional Improvement Approaches

In the "traditional way," as opposed to the "lean way," improvement efforts may fall into one of the following categories [1]:

- Problem solving = improvement
- Cyclical improvement
- Selective improvement

6.3.1.1 Problem solving = Improvement

Some organizations assume that performing the current process or producing the current product is the desired state or normal condition. As a result, an improvement effort is only needed when a problem prevents performing the work as usual. As a consequence, whenever a problem has been solved, improvement was achieved.

Fails from this approach:

1. By acting only on solving a detected problem it is both reactive (the problem is already there) and punctual (focused on solving that particular problem); it is not compatible with working to prevent potential problems and finding systemic solutions.
2. Leaving a process alone and expecting high quality, low cost, and stability is wishful thinking. Regardless how a standard is defined and disseminated, a process will tend to erode no matter what. This is not necessarily caused by the workers' poor process/standards discipline; this is the result of the process's natural interaction effects among its parts and entropy, which says that any organized process naturally tends to decline into a chaotic state if we leave it alone.
3. If a process is left alone, its performance is constantly eroding. Sometimes performance erosion is gradual, and one might suffer the "boiling frog syndrome." As a result, by the time the problem is perceived, it has already achieved great proportions.
4. Returning to the previous state (normal condition) is not really an improvement. There is no guarantee that performance will return to the previous state because the environment is constantly changing.
5. Considering items 2 and 3, one really achieves maximum normal condition performance just momentarily. This "problem solving = improvement" attitude does not really lead to any improvement at all since the company is always struggling to come back and maintain the normal condition of things (Fig. 6.3).
6. People are rewarded for fixing problems, for fire-fighting, not for analyzing, even though the problem may recur later because it was not sufficiently understood and tackled on its root causes.

Fig. 6.3 Problem
solving = improvement

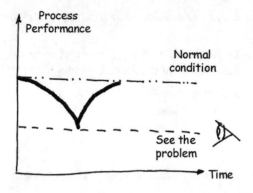

6.3.1.2 Cyclical Improvement

One "logical" alternative to avoid the complete reactiveness of the "problem
solving = improvement" attitude is to rely on cyclical/periodical improvements
and innovations. This might seem logical, if we make periodical improvement
campaigns we might tackle issues before the real problems occur; furthermore,
a campaign might create order and the opportunity to act in a systemic way.
Unfortunately, this logic is also flawed.

In order to create order, thus avoiding work flow rupture, improving is damned
until the next special effort or campaign. By doing so, this approach also conceals
a system that is static and vulnerable once the process is left by itself and eroding
between the campaigns [1].

Periodicity, even if high, is different from continuity. Depending on the periodicity (and luck as well), this approach might be capable of tackling issues before
a problem really occurs (Fig. 6.4). As a consequence, many of the improvement
campaigns ends up solving the same problems and a good part of the "improvement" effort falls on bringing back the process to a previous desired condition.

Fig. 6.4 Improvement
through periodical cycles

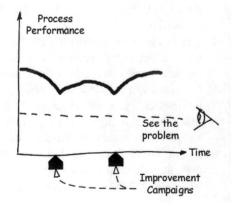

Even if the cycles' goal is to raise the process to a more desirable state than the current state, the improvement goals tend to be short term, covering only until the next planned cycle. Also, we seldom find long term goals that permeate several cycles.

By waiting for the next event, some root-cause trails might fade and the implemented improvements end up being superficial, not preventing the problems from happening again. The damned issues to be tackled during the improvement campaign can become very large and complex to tackle, preventing good solutions due to time restrictions.

6.3.1.3 Selective Improvement

The selective improvement approach is an evolution from the previous one, where a priority is given to define what issues will be tackled during an improvement campaign. Even though it seems logical and reasonable, this approach is also an unscientific and ineffective method for process improvement.

Considering the complexity of the processes themselves and of the organization performing these processes, the prioritization is mainly wishful thinking. It is like having plumbing with several leaks and believing that by fixing the main ones you would solve the entire leaking problem.

It is actually a scattershot approach, where multiple action items are initiated in the hope of both individually hitting something and of making sense as a set (Fig. 6.5).

Defining and introducing several action items simultaneously and sometimes even voting to prioritize them, indicates that we do not know what we need to do in order to consistently achieve any improvement [1]. The issues we leave out of the list will not disappear; they will grow until they become big enough to be selected to the next priority list.

Fig. 6.5 Scattershot improvements

All the previously mentioned approaches are not systemic. Considering that the PDS is a complex system, any improvement approach should have a new desired state as goal to the improvement system as a whole. Attacking problems, making punctual efforts or campaigns, and creating lists might lead to some local optimization, but will only by chance improve the system as a whole.

6.3.2 The Continuous Improvement Approach

The continuous improvement cycle is the sequence of actions to perform an improvement. Whether you call it *continuous improvement cycle*, *lean problem solving method*, *PDCA* (Plan, Do, Control, and Act), *DMAIC* (Define, Measure, Analyze, Improve, and Control), etc. its roots are in the scientific method (Table 6.1).

Table 6.1 Continuous improvement and the scientific method

PDCA	DMAIC	Scientific Method	Continuous Improvement Cycle [1]
P	D M A	1. Make observations 2. Propose a hypothesis	Pick up the problem: what is the target condition—challenge? What do we expect to be happening? Grasp the situation (what is the actual condition now?): What is actually happening? Go and see for yourself, do not learn second-hand. Investigate the causes (what problems or obstacles are now preventing you from reaching the target condition? Which one are you addressing now?): identify the direct cause of the abnormal occurrence. Use 5-Why investigation to find the chain of cause-effect relationship that lead to the root cause (see Chap. 8)
D	I	3. Design and perform an experiment to test the hypothesis	Develop and test countermeasure, which take specific action to address the root cause Rather than changing many things at once, which creates difficulties in understanding the effect from each measure taken, take one step at a time, and in rapid cycles, so you can see the correlations
C	C	4. Analyze your data to determine whether to accept or reject the hypothesis	Analyze your results. What did you learn? Do you have to dig further? What do you expect to learn by doing so?
A		5. If necessary, propose and test a new hypothesis	Follow up (when can we go see what we have learned from taking that step?) Monitor and confirm the results. Standardize successful countermeasures. Reflect on what was learned during the cycle (see Hansei events in Chap. 8) Trigger the next PCA cycle if necessary

The five questions come into play once you are striving to climb the "staircase;" you will face obstacles between each step, thus triggering the continuous improvement routine (Fig. 6.5) where the questions from Table 6.1 build upon one another [1]:

1. The better you've defined the target condition; the better your reference to assess the current condition.
2. The better you assess the current condition; the better your reference to recognize obstacles.
3. The better you recognize obstacles, the better you can plan your next step to go from the current to the target condition.

The continuous improvement operates within an overall sense of long-term direction. From day to day the improvement leads you closer to this target. Now, imagine what small effective steps of continuous improvement happening at every process, every day, can do for your company every day. Any doubts? Just look at Toyota.

6.3.2.1 Continuous Improvement Pace

Setting a target condition is like creating a "problem," and the solution is going towards it. Once the target condition acts like a beacon for all the related processes, in Lean Organization the continuous improvement and adaptation process occurs in every process (activity) and at every level of the company every day.

Day-to-day improvement happens whenever one obstacle is found, thus involving small steps. Also, changing only one thing at a time makes it easier to check the result against the expected outcome. Even if you work on several things simultaneously, refrain from changing more than one thing at any one time in a process. Such "single-factor experiments" are preferred because they let people see and understand the cause-effect relations which help develop a deeper understanding of the work process.

Remember that in complex systems, such as the PDS, the final system behavior emerges from the interaction of its parts. Changing several things simultaneously will mask what really worked from what might even have a negative impact. Therefore, no or limited learning will be gained from the system behavior.

Continuous and incremental improvement is often referred as *Kaizen*. Kaizen is a strategy where employees work together proactively to achieve regular, incremental improvements to a process. In a sense, it combines the collective talents within a company to create a powerful engine for improvement. Kaizen occurs at all levels of a company, and involves all workers related to the process under improvement. Indeed, no one understands better the process than its owners, suppliers, and customers.

Sometimes, though, incremental improvement is not enough. Imagine a radical change in the market (i.e. by substitutes) or in the work environment (i.e. due to a technological breakthrough), etc. Small incremental steps might keep the company

Table 6.2 Comparing Kaizen and Kaikaku

Kaizen Continuous incremental improvement	Kaikaku Large-scale, radical change
• Lean initiatives or events with cumulative planning and execution timelines of hours to weeks • Smaller project scope • Small to medium staff and resource allocation • Quicker results with small, individual contributions to the bottom line of the organization or value stream • Tactical	• A lean initiative or event with a planning timeline of weeks to months and an execution timeline of hours to weeks (value stream dependent) • Larger project scope • Medium to large staff and resource allocation • Results realized more slowly but with larger, concurrent, multiple contributions to the bottom line of the organization or value stream • Strategic

at an obsolete state for a very long time before it reaches the new reality/paradigm. In this case, a more fundamental larger scale and radical change is needed which we call *Kaikaku* in the lean vocabulary. Table 6.2 compares the Kaizen and Kaikaku approaches.

Kaikaku also uses the continuous improvement cycle, but solves bigger problems and requires more planning, time, and investment. As with any big change, its results take longer to fully appear because the company has to understand, learn, and adapt to the new way of doing things. After the kaikaku, kaizens will occur to help with streamlining the new process details.

6.3.3 VSMA as a Process Improvement Technique

"Doing the right product" and "doing that in the right way" are the pillars that sustain the PDP value delivery. Considering the "right product," the PDP acts like a sequence of continuous improvement cycles by itself by making incremental development cycles to guarantee that all the expected value is incorporated into the product.

For "doing it right," it works the same as all business processes within a company and can be considered a network of supplier-customer relationships where one must identify the actual value stream. The concept of value stream helps in viewing the waste and its subsequent disposal. The value stream is the set of all necessary work to bring a specific product to pass through the three critical management tasks [3]:

- **Troubleshooting** (or project flow): This includes everything from concept to product launch, through detailed design and process engineering.
- **Information Management** (or information flow in the production): This includes everything from the receipt of the order to delivery, following a detailed schedule.

- **Physical Transformation** (or flow of materials in the production): The product goes from raw material to the finished product that ends up in the hands of the customer.

Thus, an activity adds value if it creates project deliverables so that the client recognizes value in this transformation and is willing to pay for it. In this context, a product target cost becomes the minimum cost required to perform the activities that really add value, eliminating the waste throughout the process and ensuring the profit. Every task within the project should aim at creating results that are pulled by the value as requested by the stakeholders; one should be suspicious of other activities.

A typical approach for the value stream identification and improvement is the Value Stream Mapping and Analysis (VSMA), which shows all activities, inventories, and cycle times. The value stream map differs from the project schedule in that the first determines "how" things happen and the second shows "what" should be done [4].

The VSMA differs from other process improvement approaches by its ability to support the lean philosophy. The VSMA proved useful in waste elimination and by reducing the cycle-time (and related costs) to produce products that meet customer demands. A visible value chain helps to synchronize how the activities add value.

We advise to map the process and give special attention to the deliverables that are exchanged within the process. In this way, is easier to perceive the waste occurrence as presented in Fig. 5.1.

Once a waste occurrence is identified, you must go deep to identify the root causes. A very useful lean technique to support this effort is the "7 Whys," where you must ask successive "whys" until you find the root cause.

Use the following steps to waste reduction/elimination (Fig. 6.6):

1. Draw the process AS-IS.
2. Identify the process stakeholders.
3. Write down what is the value pulled by these stakeholders.
4. Identify the wastes looking at each process element as presented on Fig. 6.1.
5. Quantify the wastes impact on delivering the expected value.
6. Draw the process TO-BE (new target condition).
7. Plan and execute continuous improvement actions on each identified waste.

6.4   Continuous Product Improvement

As a consequence of continuous improvement, the PDP can be seen as a spiraling and iterative process through the PD funnel, each cycle corresponding to one PDCA round, which follow the same scientific method's steps as in Table 6.1 (Fig. 6.7), where the "C" corresponds to a particular verification or validation during the PDP.

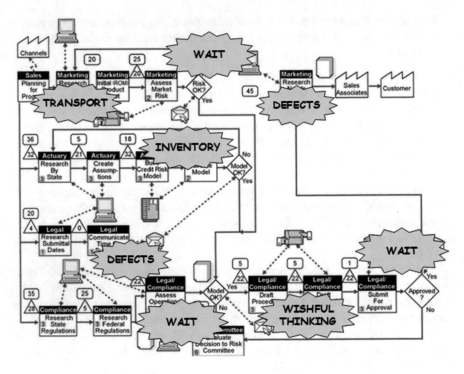

Fig. 6.6 Wastes identified in the process

Fig. 6.7 Product
development as a PDCA

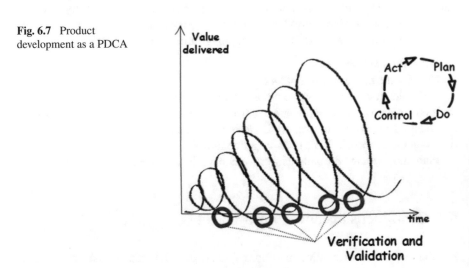

6.4.1 Point-Based Concurrent Engineering

In the traditional PDP, the development project seeks to create one alternative that
is the best (cost/risk/benefit) solution until it fails (Fig. 6.8a), for whatever reason.

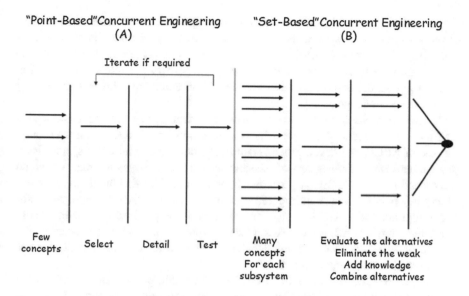

Fig. 6.8   Point-based (**a**) and set-based concurrent engineering (**b**)

This failure triggers a series of iterative loop-backs, or plan modification and resource changes. This approach is called *point-based concurrent engineering* and embeds lots of wishful thinking into the chosen punctual alternative solution [6].

Point-based concurrent engineering is based on the assumption that a good cost/risk/benefit analysis made during the PDP's conceptual phase leads to choosing the better balanced and less risky system-design alternative. Even though true, and completely logical and reasonable, depending on the product/service to be developed the residual risk might be still high, particularly after considering all the PDP particularities and performance drivers presented in Chap. 1. This is the reason why rework iterations are common in traditional PDP.

This approach also leads to a risk avoidance attitude, where the PD team discards breakthrough alternatives, which are normally high risk. In this sense, the team remains in its comfort zone, and innovation is carried out in a slow pace. Finally, the existence of a powerful advocate with a personal interest in a particular solution can bias the cost/risk/benefit analysis and jeopardize the whole development [5].

6.4.2 Set-Based Concurrent Engineering - SBCE

The lean PDP, on the other hand, does not focus on the speedy completion of individual component designs in isolation, but rather applies the Set-Based Concurrent Engineering (SBCE). SBCE recognizes that even the best alternative from the solution space have some intrinsic risk. Thus, the way to tackle this risk and avoid rework cycles is to have a "plan B" (and even C, D…).

Since the essence of simultaneous engineering is bringing downstream considerations to the table early in the development process when options are the most fluid, SBCE is a way to consider many options at this stage and then narrow the range simultaneously across functions (Fig. 6.8b). SBCE explores the solution space, supports the no-compromise attitude, allows emergent solution (combining) and creates knowledge [6–8].

In this way, SBCE is a powerful technique that guarantees the flow, avoids risk through redundancy and robustness, and allows knowledge capture. Through the use of SBCE, the development team does not establish an early system level design, but instead defines sets of possibilities for each subsystem, many of which are carried far into the design process. These sets consider all functional and manufacturing perspectives, building redundancy to risk mitigation while maintaining design flexibility. The final system design is developed through systematically combining and narrowing these sets, when alternatives are eliminated based on the growth of knowledge and confidence. The discarded alternatives are themselves considered learning opportunities.

Through SBCE, variations from an already existing product or service (incremental innovation) and completely new ideas (breakthrough innovation) can coexist. Even a solution pushed by a high level manager can be added to the set without compromising the project's progress.

Experience shows that the cost of applying SBCE is equivalent to applying the point-based approach, considering the average needed rework cycles. The great difference among them is that SBCE greatly reduces the risk of overtime, while generates more knowledge by understanding the several alternatives.

6.5 A Practical View

Considering that Product Development is a kind of knowledge-based factory where information is transformed by intellectual and non-repetitive work, seeing the waste on this process is a non-trivial task.

In practice there are several obstacles to operationalize the continuous improvement [1, 4]:

- It is hard for people to resist making a list of action items.
- The improvement cycle discipline is often difficult for senior leaders to internalize.
- People like *doing* but not *checking* and *adjusting*.
- People rather jump into solutions than making careful observation and analysis.
- The unclear path to a target condition is uncomfortable for many people. People like a clear plan in advance even though that is actually only a prediction and wishful thinking.
- Iteration (redoing steps) is uncomfortable. People feel like they did something wrong when they are asked to look again or repeat a step, yet this is very important for learning and seeing deeply.

- Many people will view improvement efforts as just another project; it seems like an improvement effort means adding more work on top of daily management duties, as opposed to it being a different way of conducting daily management.
- The big challenge for application into the PDP is the risk and uncertainty inherent to a creative and iterative process. In this case, unlike the management of most physical systems, the development activities might not result in an objective and final answer, nor is there a single correct path to achieve an optimal solution of engineering design.
- In the PDP, continuous improvement has higher impact on "doing the job right," but does not completely guarantee doing the "right job."
- In terms of the continuous improvement of the product by applying the SBCE, managers are often skeptical about investing effort in the development of several alternatives as an insurance to the PD success. There is a culture of considering rework "normal", in which is more acceptable to "do it twice" instead of "doing it right at the first time".

In order to avoid these obstacles, one must understand that improvement can be carried out either between projects or within the projects. Improvement between projects is when you consider the lessons learned from a finished development project and consider improving your standard process, so that the next project will have the benefit from this additional knowledge. The benefit of this approach is greater the more similar are the development projects, once they might have fewer differences from the standard process.

The improvement within the project uses the jidoka mindset that stops the process whenever a "defect" is identified and the team work on finding a solution for the problem's root cause. SBCE plays an important role at this moment, when the set of alternatives support a constant development progress without rework cycles.

The lean product development organization applies these two, being supported by a strong culture, knowledge management procedures and a particular organizational structure, where meritocracy is measured by supporting the value delivery through the products and/or services being developed.

References

1. Rother M (2010) Toyota Kata: Managing people for improvement, adaptiveness and superior results. McGraw-Hill, New York
2. Ohno T (1998) Toyota production system. Productivity Press, New York
3. Rother M, Shook J (1999) Learning to see: value stream mapping to add value and eliminate muda. Lean Enterprise Institute, Cambridge, MA
4. Morgan JM, Liker JK (2006) The Toyota product development system. Productivity Press, New York
5. Murman et al (2002) Lean enterprise value: insights from MIT's lean aerospace initiative. New York, NY, Polgrave
6. Kennedy MN (2003) Product development for the lean enterprise. Oaklea Press, Richmond

7. Ward AC, Liker JK, Cristiano JJ, Sobek DK (1995) The second Toyota paradox: how delaying decisions can make better cars faster. Sloan Manag Rev 43–61
8. Sobek DK, Ward AC, Liker JK (1999) Toyota's principles of set-based engineering. Sloan Manag Rev 67–83

Part III
The Wheel

A lean product development system succeeds in its objective through the Lean Product Development Process (LPD) that is itself supported by a particular organizational structure and culture and by continuous knowledge management (Fig. 1). Part III includes the elements that support the LPD. Even though the Organizational Structure is one of the wheel's elements, there is no chapter dedicated exclusively to it; we understand that the right organizational structure for you is the one that supports the lean culture and knowledge management while it is compatible with your particularities. Chapter 7 discusses the Lean Product Development Organization culture. Closing this part, Chap. 8 shows the Knowledge Management aspects of the Lean Organization.

Following the Lean Wheel System metaphor, the wheel elements presented in Part III are deeply rooted in the wheel hub elements, as well as interlaced with each other. A good wheel gives better support for the PDP, and it allows the PDP receiving full benefits from the core lean.

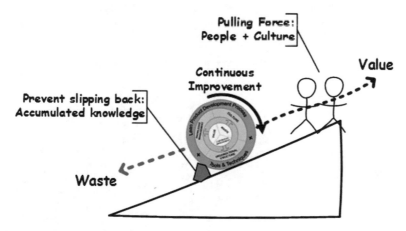

Fig. 1 Wheel elements in action

Chapter 7
The Lean Product Development Organizational Culture

This chapter presents the main aspects related to organizational cultures which support the lean philosophy. Organizational culture is unique for every organization; it is not easily defined and even harder to change. It is essentially the cause and effect of the sum of all the written and unwritten rules, attitudes, behaviors, beliefs, and traditions which contribute to the unique social and psychological environment of an organization [1, 2]. All lean processes, tools and techniques are effective if supported by a lean enabling organizational culture. This culture allows the integration and alignment from all performed processes, regardless of whether they are being supported by lean labeled or non-labeled tools and techniques. Do not forget that are not the tools used by an organization that makes it a Lean Organization, but how it uses these very tools.

7.1 Introduction

Organizational culture exists at all companies, whether it is actively maintained or left to chance. While executive leaders play a large role in defining organizational culture by their actions and leadership, all employees contribute to the organizational culture. It can be a liability or an asset once it supports the continuity of an overall corporative behavior and attitude, which is often the difference between short-term gains and long-term success. It determines [1, 2]:

- the ways the organization conducts its business;
- how the company's employees and management interact with each other, market, and the wider community;
- dress code, business hours, office setup;
- self-image, employee benefits, turnover, hiring decisions;
- the extent to which freedom is allowed in decision making, developing new ideas, and personal expression;
- how power and information flow through its hierarchy;
- how committed employees are towards collective objectives;

© Springer International Publishing AG 2017
M.V.P. Pessôa and L.G. Trabasso, *The Lean Product Design and Development Journey*, DOI 10.1007/978-3-319-46792-4_7

- guidelines on customer care and service, product quality and safety, attendance and punctuality, and concern for the environment; and
- production methods, marketing and advertising practices, and new product creation decision and practices.

Cultures develop in organizations as a result of external adaptation and internal integration and are shaped by multiple factors, including the following:

- External environment (social, political etc.)
- Industry sector
- Size and nature of the organization's workforce
- Technologies the organization uses
- The organization's history and ownership
- Metaphors, stories, rites, and ceremonies that reveal employees' shared meanings of experiences at the organization

External adaptation reflects an evolutionary approach to organizational culture and suggests that cultures develop and persist because they help an organization to survive and flourish in the particular environment it is inserted.

Internal integration is an important function since social structures are required for organizations to exist. Organizational practices are learned through socialization at the workplace. Work environments reinforce culture on a daily basis by encouraging employees to exercise cultural values.

External adaptation and internal integration are the reasons why some successful companies struggle to have the same results in different parts of the world. In reality, you will never have the same culture in all places, once external and internal factors related to each plant trigger the emergence of a slightly different culture.

Corporate culture is hard to teach in a traditional sense. It is usually learned through a defined set of corporate values, incentive systems, and ways in which people are managed, in which they communicate, and in which they prioritize. In this manner, culture is both a cause and effect of behavior throughout an organization, as well as an important enabler of high-performing companies by:

- reinforcing strategic goals of the company by aligning *what* the company does with *how* the company does it;
- supporting skill development and operational performance by fostering an environment that values learning and advancement;
- communicating the corporate brand to the marketplace through the actions of employees.

When companies succeed at integrating its values and strategy into the culture, they develop strong cultures, where staff responds to stimulus because of their alignment to organizational values. Strong cultures help firms operate like well-oiled machines, engaging in outstanding execution with only minor adjustments to existing procedures as needed. Where culture is strong, people do things because they believe it is the right thing to do.

Conversely, in weak cultures there is little alignment with organizational values, and control must be exercised through extensive procedures and bureaucracy.

7.2   Lean Enabling Organizational Culture

According to Rother [3], Toyota's culture emerged from the repetitive application and teaching of continuous improvement through 50 years. This broadly-shared cultural DNA is fundamental to the success of lean thinking and a further reason why it is a challenge, even within Toyota, to teach the lean product development system to new employees globally.

Toyota values discipline and work ethic and requires these of everyone inside and outside the company. The traditional management approach concentrates on outcome targets and consequences. In contrast, the lean approach puts considerable emphasis on how people tackle the details of a process which is what generates the outcomes. Therefore, more important than the results themselves is the logic behind achieving these results [4]. Different problems arise every day, and the solutions applied in the past might not work in this case; but the logic to achieve these solutions is a better candidate for reuse.

The unspoken business philosophy at some companies is simply to produce and sell more. Or it is about exercising rank and privilege and thus avoiding mistakes, hiding problems, and getting promoted, which become more important than performance, achievement, and continuous improvement. In this way, short-term and self-success is the rule; while long-term and group-success is the exception [3].

In this sense, the traditional company's culture assumes a clear path ahead; they hit bumps and are surprised by finding obstacles along the way and are used to firefighting. The lean way recognizes that the path ahead is unclear, focuses on going ahead with caution, and learns at every step taken. This preventive approach greatly reduces the need of firefighting and avoids its wasteful disruptive effect on the development projects' portfolio. Remember that commonly firefighting takes resources from several projects among the company, thus affecting not only the original development project, but several others that have to release resources to help putting out the fire.

By recognizing that the way ahead is unclear, the lean product development company only goes ahead after setting a vision, which acts like a beacon, pulling, guaranteeing direction, and fostering alignment to the company's actions and decisions [3].

The lean philosophy seeks the discovery of problems and respects the problems (obstacles) once their solution moves the company closer to achieving the vision. Instead of a finger pointing attitude, the lean company fosters the understanding of what is preventing the workers from having the expected performance [3, 4].

As a result, personal recognition and the meritocracy system come from technical expertise and mentoring capabilities. Only people that learned by doing and know how to communicate are capable of supporting the continuous improvement [4, 5].

Fig. 7.1 Main aspects of a lean enabler organizational culture

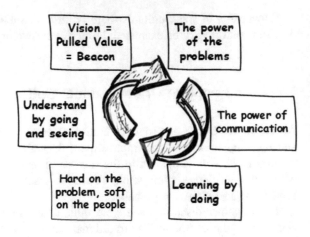

While visual and meaningful communication is needed, no one is satisfied by learning second hand. Whenever possible, one must exercise the *Genchi Genbutsu* and understanding by going to the place where the actual work is being performed, or where the problems and issues are arising, and seeing by himself/herself, and not lose any details about the real thing [4].

Figure 7.1 summarizes what we consider the main characteristics of the lean enabler organizational culture.

7.2.1 Vision = Pulled Value = Beacon

As discussed in Chap. 6, the lean organization keeps improving continuously, but no amount of effort creates real improvement if is not aligned to the defined vision and does not help the company in moving towards it. The vision represents the pulled value and is one of the continuous improvement pillars.

Indeed, no improvement can be achieved without having a final objective, and the continuous improvement process relies on first defining a new desired state. This is rather important since, on a daily basis, we are moving through unclear and unpredictable territory. Only by having the future state vision can we be sensitive to and responsive to actual conditions on the ground [3].

The vision acts like a beacon; it provides the direction in which to go, so you can always double check and adjust your course to keep moving towards your objective. Without the vision, the actions become erratic and coming and going is fairly common. You make "local improvements" but "global improvements" are only achieved by chance. The vision gives direction and provides alignment to all processes and activities.

During a product development project, the vision is represented by the sum of all valued pulled by the stakeholders through the value chain. From the project

"Would you tell me, please, which way I ought to go from here?"

"That depends a good deal on where you want to get to," said the Cat.

"I don't much care where –" said Alice.

"Then it doesn't matter which way you go," said the Cat.

"– so long as I get somewhere," Alice added as an explanation.

"Oh, you're sure to do that," said the Cat, "if you only walk long enough."

Fig. 7.2 Alice and the Cheshire cat [6]

management perspective, the vision is the project scope and all the development activities must be aligned to delivering these, and only these, results (anything besides that is not value). Any obstacle in this way must be seen as an opportunity to go further beyond the company's actual value delivery level (but this is the next topic's discussion).

Even though the need of setting a future desired state is pretty commonsense (Fig. 7.2), many times we act without having a stated and common vision of the objectives ahead. If everybody acts according to his/her understanding of how the future state should be, you have no vision at all. The vision need to be negotiated and shared with all the team so they will independently move towards it.

As a consequence, the people in the lean organization are either working on setting the vision or on achieving it.

7.2.2 The Power of the Problems

We hear about Toyota's successes but not about its thousands of small failures that occur daily which provide a basis for those successes. Recognizing that the unclear path ahead is full of problems, which themselves are opportunities to learn and create competitive advantage, is another continuous improvement pillar [3, 5]. Besides the need of a vision, recognizing the power of the problems is another cultural aspect that is a prerequisite to performing effective continuous improvement.

Toyota states that "problems are jewels" since they show us a way to the target condition. Therefore, if we have the capacity to detect problems whenever they occur (*jidoka* attitude)we expect to make faster, precise, and less risky experiments on what we need to do to keep moving forward. If there is no problem, or it is made to seem that way, then our company would, in a sense, be standing still ("no problem" = a problem). The organization should see and utilize small problems in order to exploit the potential they reveal and before they affect the external customer.

The lean company views problems as a natural part of product development; in fact, the essence of product development can be seen as a series of technical problems that must be identified and solved. From this perspective then, companies who excel at technical problem solving will do well at product development. By contrast, the traditional company's culture sees problems as negative and unexpected, an attitude that suggests problems shouldn't occur at all (Fig. 7.3). When problems do surface, as they inevitably do, there is a lot of finger pointing or blame-gaming.

A common traditional approach whenever a small problem is found is to either ignore it by "sweeping it under the rug," or logging and compiling the small problems into summaries and Pareto charts until they become relevant to be solved or wait for the next improvement cycle.

Logging problems delays the solutions. The information given by late solutions usually comes too late to be useful for process improvement efforts (the root-cause trail is already cold). Also, it is interesting to note how often the largest category in

Fig. 7.3 Blame someone else flowchart

a Pareto chart is "other," that is the accumulation of smaller problems. In the lean way, the dealing with process abnormalities should be immediate, because [3, 4]:

- If we wait to go after the causes of a problem, the trail becomes cold and problem solving becomes more difficult; we thus lose the opportunity to learn.
- If left alone, small problems accumulate and grow into large and complicated problems.
- Responding right away means we may still be able to adjust and achieve the schedule's target.
- Telling people that quality is important but not responding to problems is saying one thing but doing another and enforcing the culture on the latter.
- By having less waste, particularly inventories, lean value streams are closely coupled and a problem in one area can quickly lead to problems elsewhere.
- The situation in the process is likely to change, making it difficult, if not impossible, to understand the context where the problem first occurred.
- Because it takes so long to move forward, the pressure to solve the problem increases which causes us to jump to symptomatic countermeasures.

7.2.3 Hard on the Problem, Soft on the People

If people are threatened by problems then they will either hide them or conduct poor solving by quickly jumping to countermeasures without sufficiently analyzing and understanding the situation. The lean company fosters a proactive attitude by focusing on the solution process instead of blaming people (find me the culprit!). In a reactive company, you always need to have a quick solution; saying "I do not know" and going to see in order to understand the real problem is not an option. The lean company assumes [3]:

- People are doing their best.
- A problem is a system problem; therefore, it's reasonable to consider that it would occur regardless who is performing the work.
- There is a reason for everything and we can work together to understand the reason for a problem.

To function in this way, the continuous improvement should be depersonalized and have a positive, customer-first, challenging, no-blame feeling. In this way, at Toyota, an abnormality or a problem is generally not thought of or judged good or bad, but as an occurrence that may teach us something about our work system [4]. By being hard on the problem and soft with the people, we increase the cooperation rate and reduce the chances of problem hiding [7].

Even though at Toyota no one wastes time blaming or criticizing others, in the end, someone is responsible if something does not go right, and that someone stands up and takes the blame (or accepts responsibility) for failure [4]. This willingness to accept responsibility is the spirit of *hansei* (reflecting, identifying things that did not go well, and then taking responsibility) at work. Feeling bad and sincerely committing to doing better in the future is the driver that sustains the continuous improvement.

7.2.4 The Power of Communication

Unfortunately, many modern-day engineering managers believe their role in an organization is to attend meetings, keep abreast of the latest organizational politics, make the tough decisions about the big problems in the company, and generally look upward and outward. The philosophy seems to be that a good manager is good at delegating, and good engineers should work autonomously [3]. This approach, though, reduces the timely diffusion of information and knowledge, and its consequent reuse.

The lean way fosters (1) constant omnidirectional communication (*hourensou*), and (2) visual communication (*obeya*). In the product development environment we highlight the importance of *hourensou* and of the *obeya* [4].

Hourensou means that you must frequently report the progress of your work and its result, you must pass the actual information without your opinion, and you must ask for advice from a peer, a mentor, or a leader when you can't decide. In a cascade, the managers have the responsibility of staying informed about the activities of subordinates so they can report on key activities, give updates to their leaders, and advise subordinates.

This could be in written form, by using an email to communicate daily progress on a project to the stakeholders, or in the verbal form by making short frequent walk-ups to share and get feedback on the progress of an assignment. It is important to note the two-way nature of the communication where leaders not only listen, but give advice to the subordinates. This practice is even more powerful when combined with another important cultural trait of learn by doing, discussed later in this chapter. By "spreading the news" knowledge is disseminated [3].

At Toyota, being able to communicate effectively is essential. Assignments are given with the intent that others will follow *hourensou* from start to finish; therefore, management expects multiple updates to occur during that working period. During *hourensou*, *nemawashi* is also achieved. *Nemawashi* means preparing the roots or gain consensus. As you involve others in your assignment, you are obtaining their feedback and incorporating it so that the final product is the work of many instead of one.

Rother [3] highlights that those that have never heard of *hourensou* before may think it sounds a lot like micro-management of other's work, but it is quite the opposite. Micro-management is when you smother your team members and tell them what to do and how to do it. With *hourensou*, the team member has ownership of the assignment and makes decisions on their own, but by informing others frequently they get necessary feedback to keep the assignment progressing on schedule and within scope, both of which are key elements for reducing team member burden.

Another important aspect of communication in the lean company is to make it waste-free and visual. Visual and precise communication reduces the time for understanding and facilitates the discussions during the problem solving process.

In the product development environment, a good example of visual communication is the *obeya* (big room). The *obeya* is inspired by the military war rooms where the generals have a good sense of the current situation and strategy in order to make the best decisions. Therefore, an *obeya* contains all the needed information to position the project team for solving the project development issues. Some of the information one might expect to find in an *obeya* are:

- Project vision
- Project scope
- Schedule
- Indicators
- Risks and assumptions
- Project team and roles
- Current state of the product (design, prototype, etc.)
- Trade off curves of interest
- Any other useful information

As expected, the type of information to be presented in the *obeya* may differ from project to project, and from different phases of the same project. The *obeya* can vary from very simple (Fig. 7.4) to complex.

In order to facilitate the go-and-see and support better meetings and decisions, the *obeya* might include a product prototype, or even be moved closer to the place where the actual product is (particularly in the case of large products, such as buildings, aircrafts, ships etc.). It is also dynamic, since its contents evolve to better support the development team at each product development phase.

7.2.5 Learning by Doing

In a lean system, people learn best from a combination of direct experience and mentoring [3–5, 8]. Classroom training and simulations cannot ensure change, mastery, and consistency. This learning-by-doing way of thinking is part of the lean DNA. It should be in employees' genes to try out options and learn from actual experiences. Leaders are therefore teachers, encouraging and watching for the right opportunities to impart significant lessons.

Part of this is knowing how to deal with the people being taught [4]:

- Don't lay off people at the first sign of each business downturn.
- Don't pit people against each other so you can reward the winners and turn off the losers.
- Don't leave new employees to their own devices or ambitions to learn on their own.

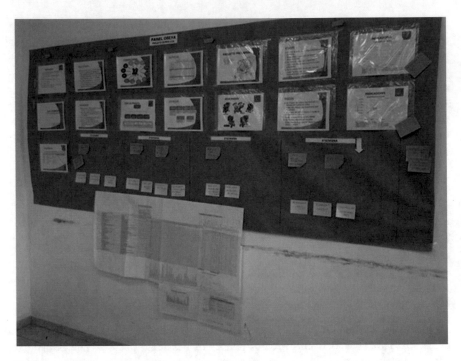

Fig. 7.4   A simple *obeya* (photo by the authors)

Consequently, the primary task of managers and leaders revolves around increasing people's knowledge and capability of performing the improvement routine, and mastering this task is part of their evaluation, bonus, promotion, and salary.

In this learning system, each employee is assigned a more experienced employee—a mentor—who provides active guidance through the process of making actual improvements or dealing with work-related situations. That mentor, in turn, has his or her own mentor who is doing the same. These mentor-mentee relationships though are not necessarily linked to the organization hierarchy.[1]

The mentor-mentee dialogue often begins with the mentor giving the mentee a purposely vague assignment, need, or challenge. The mentor then asks what the mentee proposes. The mentee's answer helps the mentor discern how the mentee is thinking and what input should the mentor give next. The mentor asks "why." The mentee's role becomes planning and carrying out improvement cycles, also with oversight of the mentor [3].

The mentee is the person who works on the problem whereas the mentor's task is to keep the mentee "in the corridor" of the continuous improvement routine. While the mentee is responsible for doing, the mentor bears considerable responsibility for the results but should not give solutions to the mentee. The mentor lets

---

[1]See [3] for an in depth Toyota's mentor-mentee system description.

the mentee make small missteps, as long as they do not affect the customer, rather than giving the mentee answers up front. This overlap of responsibility creates a bond between mentor and mentee because if a mentee fails then it is the mentor who will get the scrutiny.

Although the mentor is often a tough customer who leads the mentee through the problem solving via questioning, ultimately the mentee is the person who must analyze the problem and develop the countermeasure. If the mentee sufficiently solves the problem in a way that meets the target condition, then the mentor must accept this.

Note that the goal is not necessarily to develop the very best solution today, but to develop the capability of the people in the organization to solve problems. The mentor gests no extra reward for having a better idea than the mentee. The mentee's solution though must be good enough to serve the customer, but beyond that, having the most perfect solution is not what they want [3].

Indeed, the mentee's performance reflects the current capability of the organization. Once the solutions the mentees develop reflect the current level of capability in the organization, they can be an important input for mentors. Artificially creating perfect solutions would disguise the true state of affairs and make it more difficult to understand what we need to do next to move our organization forward.

A mentor must have the right mindset of going and seeing. Mentees often feel the pressure to give an answer, even if they do not have sufficient basis to remove all wishful thinking. The mentor should get himself and the mentee to the point where "I don't know" is an acceptable and valid answer. And when you say "I don't know" you should then go and see!

When you go and see, you shall be open-minded, neither having preconceived notions about what could be the situation, nor about the possible solutions. The mentors should know very well how the continuous improvement proceeds (the how), but should have an open mind in regard to the content of the particular improvement effort (the what).

Inexperienced mentors often ask questions directing the mentee to adopt the mentor's preconceived solution. At Toyota, you have to be a mentee before you can mentor and in order to become a mentor you must have sufficient experience in carrying out continuous improvement [3]. A good parallel can be made comparing this system to martial arts when you are only promoted to the next belt when you have mastered all the movements from the current one.

Therefore, Toyota engineers have a career path based on demonstrated competence where the managers have been developed through the same mentor-mentee system and usually know a job better than the engineers reporting to them do. As a result, the learning by doing works extremely well as it is perpetuated across generations of engineers [3–5].

Besides rewarding the technical competence, all the LPDO's organizational structure is aligned by the common objective of delivering value to the customer. Therefore, at the end of the day, all the development functions are measured by the value added through the value chain.

7.2.6 Understand by Going and Seeing

Managing from a distance through reported metrics leads to overlooking or obscuring small problems, but it is precisely those small problems that show us the way forward. Overlooking or obscuring small problems inhibits our ability to learn from them while they are still understandable, and to make timely adaptations in small steps [3, 4, 8].

In order to avoid that, Toyota practices the going to the source (*Genchi Genbutsu*), where you go and see the actual situation first hand, which promotes deep understanding of the current reality. According to Kiichiro Toyoda, "one can never trust an engineer who does not have to wash his hands before eating dinner." [4, 8].

The main point of *Genchi Genbutsu* is that you can only develop quality products by having your engineers intellectually, physically, and emotionally connected to those products. Some ways that Toyota practices it are [4]:

- **Value targeting process**: Understand what each stakeholder values; provide deep understanding of the stakeholders, particularly the customers (check the Lexus case box next in this section).
- **Product use analysis**: Understand how the value expected meshes with the program's product performance and characteristics; provide deep understanding of the customer experience/expectations with/from the product (check the Toyota Sienna case box next in this section).
- **Prototype builds**: Participate in both virtual and physical prototype builds. When engineering changes are deemed necessary during the prototype phase, they are often made on the spot where issues are identified.
- **Daily build wrap-up meetings**: These are attended by the product development team, including the suppliers at the end of each day. The meetings are held right at the building site were participants can witness firsthand the quality, cost, productivity/ergonomic, or any other issue, and where they record issues/countermeasures and give new assignments on the spot.

During the mentor-mentee process, where going and seeing also keeps the mentor closer to the real condition in the process, if you rely on reports alone, rather than going to see for yourself, you will quickly not be able to give good advice.

THE LEXUS CASE[2]
In 1983, Toyota chairman Eiji Toyoda summoned a secret meeting of company executives to whom he posed the question, "Can we create a luxury vehicle to challenge the world's best?" This question prompted Toyota to embark on a top-secret project, code-named F1 ("Flagship One") (Fig. 7.5).

---

[2]Adapted from http://www.toyota-global.com/company/toyota_traditions/innovation/jul_aug_2003.html + Relentless pursuit.

Fig. 7.5 A luxury vehicle to change the world's best

The F1 project, whose finished product was ultimately the Lexus LS 400, aimed to develop a flagship sedan that would expand Toyota's product line, giving it a foothold in the premium segment and offering both longtime and new customers an upmarket product.

A design study team went to the United States spending time in focus groups and with dealers, getting to know the customer. Going above and beyond the usual process with eight presentations over a period of 16 months, designers and management went back and forth until May 1987 when the final design was approved. During that time, several F1 designers rented a home in Laguna Beach, California to observe the lifestyles and tastes of American upper class consumers. Meanwhile, F1 engineering teams conducted prototype testing on locations ranging from the German autobahn to U.S. roads. Toyota's market research concluded that a separate brand and sales channel were needed to present its new flagship sedan and plans were made to develop a new network of dealerships in the U.S. market.

The TOYOTA Sienna CASE[3]

Yuji Yokoya, a Toyota engineer, was given responsibility for re-engineering a new generation of the Toyota Sienna minivan for the North American market. So he drove one more than 53,000 miles across America, from Anchorage to the Mexican border and from Florida to California (Fig. 7.6).

Crossing the Mississippi River by bridge, he [Yokoya] noted that the Sienna's crosswind stability needed improvement.

He observed excessive steering drift while traversing gravel roads in Alaska and the need for a tighter turning radius along the crowded streets in Santa Fe.

Driving through Glacier National Park, he decided the handling needed to be crisper. He also made an all-wheel-drive option a priority along with more interior space and cargo flexibility.

---

[3]Adapted from http://www.economist.com/node/14299017.

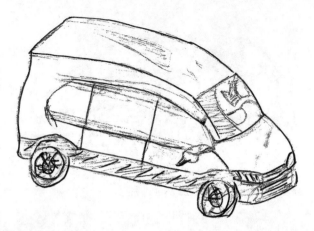

Fig. 7.6 A minivan for the whole family

Finally, he decided that the new Sienna would have to be a minivan that families, and especially kids, could live in for extended periods of time. Upgrading seat quality became a priority, along with "kid friendly" features such as a roll-down window for second-row passengers, an optional DVD entertainment center and a conversation mirror so parents could monitor what was going on in the back seat.

"The parents and grandparents may own the minivan," Yokoya said, "but it's the kids who rule it. It's the kids who occupy the rear two-thirds of the vehicle, and are the most appreciative of their environment."

 7.3 A Practical View

As a rule of thumb, cultural change will never be really achieved without the high management commitment. Thinking about the development organization, some initiatives are impossible to take if the development group/department is trying to sail in a different direction from the rest of the company. This is particularly true considering anything that has to do with meritocracy.

Considering the six aspects mentioned in this chapter, practicing "go and see" and using some of the "power of communication" can be exercised in an individual way, particularly if you are responsible for a team. Even though the chances of having even these initiatives spread without the high management support are rather low, to do nothing is not an option for the true lean advocate (Fig. 7.7).

Fig. 7.7 To do nothing is not
an option

What if we don't change at all...
and something magical just happens?

References

1. Schein EH (2010) Organizational culture and leadership, 4th edn. San Francisco, Jossey-Bass
2. Ehrhart MG, Schneider B, Macey WH (2013) Organizational climate and culture: an introduction to theory, research, and practice. Routledge, New York
3. Rother M (2010) Toyota kata: managing people for improvement, adaptiveness and superior results. McGraw-Hill, New York
4. Morgan JM, Liker JK (2006) The Toyota product development system. Productivity Press, New York
5. Ward A (2007) Lean product and process development. The Lean enterprise Institute, Cambridge, MA
6. Carrol L (1865) Alice's adventures in the wonderland. Macmillan & Co, Chicago
7. Fischer R, Ury W (2011) Getting to yes: negotiating agreement without giving in. Penguin Books
8. Womack JP, Joncs DT (2003) Lean thinking. Free Press, New York

Chapter 8
The Lean Product Development Organization Knowledge Management

Organizational learning is one of Toyota's core competitive advantages [1], as a consequence Toyota's organizational structure makes it a true leaning organization. This chapter discusses the main aspects related to knowledge management in the lean organization context. Once the lean product development process (as any process that embeds the lean philosophy) shall support knowledge management, here we present some techniques related to knowledge identification, creation, dissemination and use. The LPDO not only learns from itself, but from the customers, the suppliers and the competitors. Useful knowledge, both from the product and from the process, is represented in a way to facilitate not only its future use and dissemination, but also the knowledge management and evolution. Remember that more important than the concepts themselves is the reason and value behind them.

8.1 Introduction

The lean organization culture, as presented in the previous chapter, has the potential of creating a true learning organization. In fact, anyone familiar with lean thinking understands that organizational learning is one of Toyota's core competitive advantages [1]. What is less understood is that organizational learning is only possible with living standards that are seriously followed and regularly updated [2]. Using standards as straightjackets is ineffective and not lean; they should be used as a reference and represent a desirable target condition to be achieved.

In many companies the accumulated knowledge is used to specify and post standards, and by doing so, believe that they have established discipline, accountability, or control of the workers. Posting a standard, though, is not the end, but just the beginning; having a standard is having a reference point to make a planned versus actual comparison possible so that gaps between what is expected and what is actually occurring become apparent. In this way, the company indeed uses the accumulated knowledge to define the standards, which are indeed the defined

© Springer International Publishing AG 2017
M.V.P. Pessôa and L.G. Trabasso, *The Lean Product Design and Development Journey*, DOI 10.1007/978-3-319-46792-4_8

vision, from which it can see what the true problems are, where improvement is needed and keep learning from these improvement cycles.

Rother [3] states that rather than asking "Have we posted work standards?" we should ask "How do we achieve standardized work?" While the traditional company says it has standardized work when it has standards posted to all involved parties, the lean company checks to see if the observed process matches the standard. If there is a difference between the two, and there often is, they say, "Not yet" (note that "not yet" differs from "no" since it implies motion—on the way). These two approaches parallel to the traditional improvement approaches and the lean continuous improvement approach, as presented in Chap. 6.

A company that cannot standardize work struggles to learn from experience, and is not truly engaged in lean thinking. Indeed, any company that simply tries new things without standardizing is "randomly wandering through a maze," repeating the same errors.

Toyota's quality excellence does not result from repetitive processes, but from strivings to achieve the target condition of the process being done the same way each time [3]. The difference is subtle, but it's important if you want to succeed in your lean journey.

Indeed, the path from posting the standard and actually having standardized work is when actual learning occurs. Therefore, in order to support the value delivery while reducing waste, unevenness and overburden, the Lean Product Development Organization (LPDO) must be a true learning organization, transforming data into wisdom in the most efficient way (Fig. 8.1). In fact, through the lean organization culture and applying the knowledge management (KM) practices, the LPDO aims to create standardized (desired state) collective wisdom.

Continuous improvement is the motor that supports the climbing from understanding relations to understanding principles. The accumulated knowledge is materialized into standards, which are updated at each further step. In this way, standards represent the company's current knowledge state [2, 3].

Product development, more than any other part within the value stream, is where the most effective use of people is required. Interactions between individuals or groups tend to be non-linear and often unstructured and thus difficult to see. Unlike design drawings, which can be inspected and verified, it is virtually

Fig. 8.1 From data to wisdom

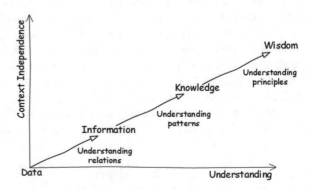

impossible to map the ways in which intellectual capital is applied during the development of these same designs. This scenario makes it even more challenging to define standards and apply knowledge management in order to support achieving standardized work.

Knowledge Management involves setting the right organizational structure and responsibilities and comprises a range of practices, tools, and techniques used to identify, create, represent, distribute, and enable adoption of tacit and explicit knowledge. In PD, such knowledge can be related to either the product itself or the development process. Some of the practices, tools, and techniques applied by Toyota are presented in sequence. The reader must remain aware that the real value lies on the reason behind using them, rather than on the practices, tools and techniques themselves. Therefore, one always must ask why and understand the value they deliver.

8.2   Organized to Learn (and to Lean)

Organizational structure and how responsibilities are defined have a big impact on the smooth running of the PDP and on knowledge creation and use. This organization promotes effective communication within the LPDO which leverages the benefits from the PD team's multi-expertise, multicultural, and multidisciplinary characteristics [4].

On each product development center (Fig. 8.2), Toyota uses a weak matrix structure, where the chief engineers rely on the center's head support to deal with the different functional divisions [5]. While in most companies with a matrix, engineers have conflicting allegiance to the functional boss and the program

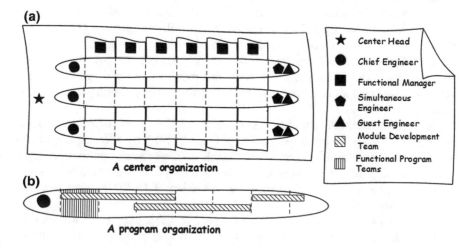

Fig. 8.2 The PD center organization (**a**) and the PD program organization (**b**). Adapted from Cusumano and Nobeoka [5]

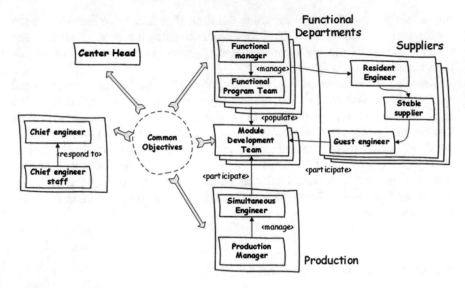

Fig. 8.3 The development organization roles

manager, Toyota uses customer pulled value metrics and incentives to direct engineers toward common objectives (Fig. 8.3).

In the development center organization, the center heads are also responsible for coordinating all the different vehicle projects within the center. Because both project and functional divisions report to one manager, this structure should reduce conflict.

At any one point, there are thousands of Toyota associates working on a program, but the chief engineer has a staff of only six to ten people who formally respond to him. The chief engineer has no direct ascendance over the rest of the team or his/her staff; they depend on the functional managers to provide people from their functional teams to development program's module development teams (MDT) [2]. Similarly, the production department assigns simultaneous engineers and even guest engineers from the suppliers are welcomed to complete the team. This arrangement creates real concurrent engineering and provides synchronization.

It's important to highlight how the lean model relies on the suppliers. Indeed, the suppliers are an important part of the extended LPDO by allowing the LPDO to concentrate its own resources in the defined strategic areas and by creating a capacity cushion that supports the LPDO on the absorption of demand fluctuation. In fact, Toyota considers each supplier an extension of its PD process and lean logistics chain [2, 5].

Toyota generally works with a small number of stable suppliers for each outsourced part (typically two or three). Many times they use the keyretsu model where both companies hold equity in each other.

Even though the suppliers play an important role in the macro development organization, Toyota is not willing to give up internal competency even though it may be cheaper or more convenient to do so. Moreover, when Toyota outsources, it does not relinquish control; it wants to learn and excel in new technology along with suppliers [2].

Some definitions of the roles shown in Fig. 8.2, as adapted from Morgan and Liker [2] are:

Chief engineer is a prestigious position in the company, being accountable for project's results. The chief engineer and the **chief engineer staff**, through deep knowledge of the customer's needs, represent the voice of the customer and give common vision/goal to all functional teams involved in the PD. The range of responsibilities for the chief engineer and his or her small staff includes defining the product concept, the program objectives, the product-level architecture, performance, characteristics, and timing.

Functional program teams are the functional divisions related to the development program itself. They are technical specialty groups with their own **functional managers** who supervise the engineers and decide which project they are assigned to, conduct their performance evaluations, and determine promotions.

Module development teams (MDT) are the cross-functional teams responsible for each product subsystem. They identify and resolve technical problems and map strategies to achieve component/subsystem level goals that are aligned with and support overall product objectives. To avoid interrupting the flow as a new product moves from one organization or resource to another, the cross-functional module-development team must perform concurrent engineering, thus, synchronizing individual functional organizational activities. Effective cross-functional synchronization in a lean PD system requires a thorough understanding of: (1) the details of how the work actually gets done; (2) each participant's specific roles and responsibilities; (3) key inputs, outputs and interdependencies for each activity; and (4) sequences of activities in all functions

Simultaneous engineers promote a stronger and more intensive involvement between product engineers and production engineers at a higher level which is needed to coordinate the extra complexity and need for speed. Some key production engineers may be assigned to MDTs and function as full-time representatives of their manufacturing disciplines. The simultaneous engineers are also responsible for hitting both investment and variable cost targets for their parts—both tooling and the parts produced by these tools—as set by concurrent engineering. He is responsible for his parts until the start of production. In the preparation for the *kentou* (see Chap. 9), the simultaneous engineers spend a great time on the production plants gathering data and talking to team leaders and operators in order to understand fully current manufacturing issues and solicit potential countermeasures.

Guest engineers are engineers from suppliers who reside full time in the LPDO product development office. While they have separate areas, they interact daily with the LPDO engineers. This has an obvious benefit—free engineer resources for Toyota. But that is not the purpose. The purpose is integration. When the LPDO invites a supplier to send guest engineers, it is a significant commitment to long term co-prosperity.

Resident Engineers are engineers exchanged on temporary assignment both within Toyota and with affiliated companies. This is a learning opportunity for the resident engineer and is also a method of standardizing practices and processes between Toyota and its suppliers.

We do not advocate that you need to create an organizational structure identical to Toyota's. We, though, emphasize the way the structured the organizational roles and even the suppliers around the pulled value and in a way to foster knowledge management. Indeed the pulled value is the vision towards all the elements from the LPDO should work in order to deliver the right product and or service.

8.3 Knowledge Identification and Creation

The great challenge to knowledge identification and creation is to recognize which pieces of knowledge must be kept. The primary learning source of the LPDO is itself, but the company also learns from the customers, the suppliers, and the competitors. In sequence, a series of practices is presented. Your company might use different practices while delivering the same value (adapted from [1, 2]):

5 Whys: Toyota has a practice of asking why five times to solve problems at the root cause. It is an iterative interrogative technique repeating the question "Why?" Each question forms the basis of the next question. The "5" in the name derives from an empirical observation on the number of iterations typically required to resolve the problem.

A3 Process: The A3 process is a Toyota-pioneered practice of getting a problem, an analysis, a corrective action, and an action plan written down on a single sheet of large paper, often with the use of simple graphics. The A3, described further in this chapter, is a powerful management, learning, and continuous improvement technique.

Communities of practice: The several functional areas and program managers from various projects have their specific communities of practice to discuss lessons learned and to pass on new standards. The lessons are derived from each program's *hansei* events that he/she takes part.

Competitor Teardown and Analysis: Teardown exercises provide an opportunity to learn about competitors. The benchmarking is owned by a team of engineers who specialized in and were responsible for the analyzed subsystem, module, or part. While benchmarking, the team defines specific problems and work on countermeasures. The same group that did the benchmark is the one to implement the changes on the company's own product. Ownership, responsibility, and good problem solving are all keys for a successful lean PD process. This hands-on exercise is another example of *genchi genbutsu* and an excellent way for engineers to learn.

Cross-checking: This is one method to discover problems and check quality, especially from the prototype phase onward. This applies to the process of understanding the true condition of parts and the appropriateness and accuracy of the measurement system that is being employed. You can achieve cross-checking by requiring several groups to check the same parts/data independently.

Daily wrap-up meetings: This is another potent learning and problem-solving mechanism utilized during design reviews, prototype builds (physical and virtual), tool manufacture, and launch. Held at the end of each day, typically on the shop floor where the work is being done, the wrap-up meeting is attended by all key participants, including suppliers. It clarifies assignments, and generally aids in real-time, course-correction decisions. Furthermore, the wrap-up meeting is a strategy that captures lessons learned. Alternatively, the team could make "also/ or meetings" at the beginning of the day, where they discuss the issues from the previous day, define solving strategies, and set the strategy for the day. Both these meetings are good opportunities for *hansei*.

***Hansei* events**: *Hansei* is a Japanese word for reflection. At these reflection events, participants share their PD program experiences, lessons learned, project shortcomings, identify things that did not go well, take responsibility, and then discuss and develop countermeasures.

Ijiwaru: Testing and validation can be another important opportunity to learn from experience. In most companies, required performance specifications are set in advance, and designs are tested for compliance to these specifications. Learning in this environment is minimal because it is strictly a pass-fail metric. *Ijiwaru* testing is the practice of testing subsystems to the point of failure. By testing these subsystems under both normal and abnormal conditions and pushing designs to the point of failure, the engineers gain a great deal of insight into both current and future designs and materials by understanding the absolute physical limitations of their subsystems. This practice also gives a great deal of confidence in the performance parameters of their products in the hands of the customers, and is a key to producing good trade-off curves (described later in this chapter).

Problem solving at the source: In PD, it is crucial to solve problems early, at the source and permanently, and to learn from these problems in order to improve the organization. The standardized scientific problem-solving process (see Chap. 6): identifies the problem's root cause, evaluates the potential impact of several possible solutions, and produces a high quality countermeasure that can resolve the immediate issue as well as prevent its recurrence. Subsequent *kaizen* and *hansei* events verify the countermeasure and the results can be communicated across programs by updating standards and checklists which are increasingly part of the "know-how database."

Product use analysis: This analysis aids in understanding how the value expected meshes with the program's vehicle performance and characteristics and provides deep understanding of the customer experience/expectations with/from the product (example in Chap. 7).

Rapid learning cycles: In companies with very slow-moving development programs and frequent job rotation, engineers rarely have the opportunity to experience more than one development program so they focus only on one aspect of the product. Within a product development program, each major phase is a mini-cycle of the PDCA model; the entire PD program is a macro-level reflection of the cycle. The faster the product development cycle, the more can be cycled through it. Most importantly, PDCA develops towering technical competence and supports continuous learning.

Supplier technology demonstrations: At the beginning of each program, suppliers demonstrate technology that might be appropriate for the new product by bringing parts and meeting face to face with the LPDO engineers. This is a good opportunity for the LPDO engineers to learn about new developments and for the LPDO to leverage supplier resources fully.

Value targeting process: The result provides understanding about what each stakeholder values and provides deep understanding of the stakeholders, particularly the customers (example in Chap. 7).

8.4 Knowledge Representation, Distribution, and Enabling

Useful knowledge, both from the product and from the process, is represented in a way to facilitate not only its future use and dissemination, but also the knowledge management and evolution.

A good knowledge representation is paramount to make its future use waste-free and even possible. Imagine if you kept the original reports from all performed tests on the products ever produced by your company. Sorting useful information about some specific variables of interest by revising all this documentation would be unpractical, wasteful, and might prevent anyone from trying to do so. On the other hand, if, instead of storing raw data in your knowledge base, you keep aggregated and graphical information and/or sorted best practices from what has worked in past development projects, future reuse and evolution of this information would become easier and valuable.

Some practices to store the information in such a useful way are (adapted from [1, 2]):

Trade-off curves: A trade-off curve is a relatively simple tool that is consistently used by Toyota engineers to understand the relationship of various design characteristics to each other. In a trade-off curve, a subsystem's performance on one characteristic is mapped on the X-axis while the other is mapped on the Y-axis. A curve is then plotted to illustrate subsystem performance relative to the two characteristics (Fig. 8.4). Trade-off curves might be used to evaluate speed to fuel economy in the tuning of a given power train configuration, or the size of a radiator to its cooling capacity. A particular MDT or supplier can, for instance, create several different prototypes for one particular subsystem, module, or piece. By making many prototypes the team can vary different factors and make tests to develop trade-off curves so that the chief engineer could understand the relationship of particular aspects of interest. Trade-off curves are a fast and effective

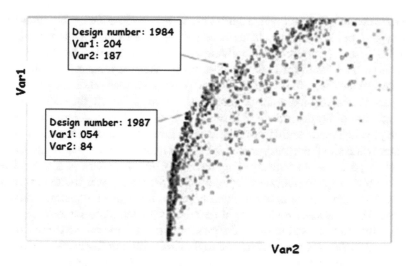

Fig. 8.4 A trade-off curve example

way to communicate very complex and technical performance attributes during the development of multiple alternatives. Some of the trade-off curves attributes are: simplicity, knowledge capture, experience sharing, communication enterprise wide, and shortened technical learning curves.

Common critical aspects on the product geometry: Communizing certain shapes, forms, and holes for efficient manufacturing, allows: re-use, reliability, efficient manufacturing, communication enterprise wide, and shorter technical learning. This practice is closely related to some DFM guidelines such as minimize part variation and emphasize standardization.

Common product architecture-standard aspects: When there is common architecture application through detailed design standards and specifications the engineer can draw from a database. The engineer can expand, shrink, or otherwise modify these structural best practices while the database simultaneously maintains critical geometric relationships to preserve product performance and manufacturability. Whenever possible, the engineer identifies carry-over or cross-platform parts for possible re-use. Common construction sections are a standardization tool used to capture standard architecture for each part and provide design anchors for each product. They dramatically reduce the amount of work required from the engineers as new design styles are considered. The content of this knowledge representation practice is the very core of the KBE, Knowledge Based Engineering explained and exemplified in Chap. 2.

Common product platforms: This is a collection of the common elements, especially the underlying core technology, implemented across a family of products. Some of the common product platforms advantages are: re-use, reliability, safety, flexibility, addressing critical issues on platforms, and development speed.

Product design standardization: This is standardization of product/component design and architecture. It includes the use of proven, standard components shared across vehicle models, building new variations on common platforms, modularity, and design for (lean) manufacturing standards that create robust, reusable, design architecture. Many of Toyota's design standards are not given as specific parameters requirements or directives; more typically, they are concerned with ratios and physics driven. These are sort of "if, then" statements based on proven physical realities that give engineers a great degree of latitude and creative freedom while simultaneously maintaining lean manufacturing requirements.

Design patterns: In software engineering, a design pattern is a general reusable solution to a commonly occurring problem within a given context in software design. A design pattern is not a finished design that can be transformed directly into source or machine code. It is a description or template for how to solve a problem that can be used in many different situations. Patterns are formalized best practices that the programmer can use to solve common problems when designing an application or system. Design patterns can speed up the development process by providing tested, proven, development paradigms. Effective software design requires considering issues that may not become visible until later in the implementation. Reusing design patterns helps to prevent subtle issues that can cause major problems and improve code readability for coders and architects familiar with the patterns. This practice is one example of Design for Modularization.

Engineering skill-set standardization: This is the standardization of skills and capabilities across engineering and technical teams. A new engineer's career path consists of experiences that develop deep technical competence while slowly climbing the technical hierarchy within each functional department, and is a direct result of engineers being rewarded for technical achievement. The engineer's boss usually knows how to do the job better than the engineer; he or she also knows the standardized process for doing it, which enables the leadership principle of teaching and mentoring. The lean PD system depends on mentoring for developing talent. To support the mentor/mentee system, Toyota creates an engineering apprenticeship environment in which highly technical tacit skills are handed down from one generation to the next, thus basing professional growth on demonstrated competence in the real world.

Process logic: Process logic determines who will do what and when, and which decisions the PD teams must make at each milestone in the product development process at macro level; it makes no attempt to provide all the details or how the work is done, but it does provide the framework that coordinates all the various participants. The functional organization that fully understands the process creates, maintains, and owns the detailed work instructions. Process logic by itself cannot create flow, but when it is flawed, it drives rework loops, waste, and prevents flow from taking place. Toyota's approach to macro-level process logic is the essence of elegant simplicity. It provides centralized control without the waste associated with monstrously large traditional PD central schedules (which are usually too complex to follow accurately) and places ownership and accountability where it belongs.

Process standardization: One level lower than the process logic, process standardization involves standardizing tasks, work instructions, and the sequences of tasks in the development process itself. This category of standardization also includes the downstream processes of testing and manufacturing the product. It enables true concurrent engineering and provides a structure for synchronizing cross-functional processes. A standardized development process means standardizing common tasks, sequence of tasks, and task durations, and utilizing this as the basis for continuous product development process improvement.

Engineering checklists: These are simple reminders of things that should not be left out. Ideally, engineering checklists are an accumulated knowledge base reflecting what a company has learned over time about good and bad design practices, performance requirements, and critical design interfaces that are critical to quality characteristics, manufacturing requirements, and standards that communize design. Checklists may define crucial steps within a process (process checklist) or provide guidelines for specific characteristics of a product design (product checklist). They are based on firsthand experience and are updated and validated regularly to incorporate any new or technological developments. In all cases, these checklists contain very detailed information about the product or process. Furthermore, the same groups that use the checklists maintain and update them at the end of each program and, as required, at *hansei* events. In the LPDO, maintaining the checklists is never a corporate IT function, or the amorphous responsibility of "engineers."

Know-how databases: The computerized know-how database is the collection of standards combined with design data and tools such as digital assembly. The functional organizations that use these databases maintain, validate, and update them as needed. Indeed, the know-how databases evolved from the engineering checklists, which worked very effectively in paper form before they were ever computerized. The notebooks were not profound and did not replace deep engineering knowledge; they just reminded the engineer to think of each aspect: Did you check whether two parts are interfering with each other or not? Did you check that the gap conforms to standards? Does this ratio fall within a standard range? Did you apply DFX or any other recommended technique? There may be a graph showing not to exceed a threshold. In each case, the engineer physically makes a check to note, "I thought of that" or, "I did that." It is much like a pilot's flight checklist—it does not make the pilot a great pilot, but it can help avoid basic mistakes. One potentially big step forward in computerizing the checklists is that they have developed from simple rules of what to avoid or what numerical values to use into explanations of the reasoning behind the rule. An engineer who saw the old checklist had no way of knowing the reason behind the rule. It provided know-*what* but not know-*why*.

8.4.1 A3 Report Planning Method

The A3 Report is a method of presenting a story in a one-page document. The "A3" comes from an international size for a sheet of paper with 29.7 cm per

42.0 cm. The report format generally mirrors the continuous improvement steps. It is written in a succinct, bulleted, and visual style that tells a story with data. Although the A3 is typically on one page, there can be additional pages of backup data. It is the "story" itself that is built up and presented on the single page.

The A3 [2]:

- enhances logical thinking and decision making (when to do/when not to do);
- provides a standardized method of communication;
- supports quick decision making to more easily spot errors;
- facilitates cross-functional management;
- facilitates cross-cultural communication;
- facilitates the analysis of the solution space and the knowledge identification and creation, and
- focuses on problem-solving activities.

There is no magic in the A3 documents themselves. The value lies in the process of making and using it. Having a filled and signed A3 is just a formality. Most of the benefit of an A3 lies in the process of creating it, because it forces you to work with facts and data and think through what you are doing [6]. The process of boiling a project down to the essential facts and creating a visual one-page report is excruciating. Having the document being slid back and forth between mentor and mentee several times progressively develops better understanding. Americans who work for Toyota report that this is one of the most difficult and at times frustrating processes to learn. You must have disciplined workers who have an absolute commitment to the process, no matter how uncomfortable and onerous that process is.

There are four types of A3 stories, namely [2]:

1. **Proposal story**: The proposal story creates a plan when a new direction and policy is made, or there is a company value or policy that is not being addressed or needs to be changed.
2. **Info story**: The info story conveys general information to any audience, inside or outside the company; it only summarizes the current situation and does not include an evaluation component.
3. **Status story**: The status story reports the current situation of an ongoing plan.
4. **Problem solving story**: This story is used when a plan, goal, or standard exists but is not being met.

Some of the vital points for a successful A3 report creation are [1, 6]:

1. Plan time to grasp the complete situation.

   - Consider a wide range of information sources.
   - Consider others involved, lay the groundwork, and build consensus (*nemawashi*).
   - Base story on facts, not opinions and data alone.
   - Consider the long term effects.

2. Decide what kind of story you need to tell. Write the story to your audience in order to fulfil their needs and increase their knowledge of the situation.
3. Root the story to the company's values and philosophy.
4. Make your story flow in a logical and concise sequence.
5. Save words by using graphs and visuals whenever possible, and clarify the accuracy of data used.
6. Make every word count, be specific, avoid specialized language.
7. Consider the visual effect of each box on the page in helping you tell the story.

Some caution has to be taken, though, in the sense that a written document can encourage e-mail communication over face-to-face communication, or be used as a substitute for Go and See. Communications should remain face to face and you should seek facts over data at the process.

8.4.2 Problem-Solving A3

This kind of A3 embeds the PDCA (Fig. 8.5), where fields I to IV relate to the effort of grasping the situation through Go and See, and fields V, VI, and VII support the PDCA Do, Check, and Act activities, respectively.

I.   **Background**: Answer why you are going to develop this particular product. Explain the vision and why achieving it makes sense. Root the vision to the company strategy and the market opportunities. Consider analyzing the market trends through the years and how they evolved in response to relevant scenarios (economic, political, demographical, etc.).
II. **Current Condition**: Explain the current situation, in the sense of the current stakeholders, their pulled value, and how your products (if any) and the competition are performing at delivering this value.
III. **Target/Goals**. Once you understand the background, its trends, and the current product/market conditions, you can set some targets/goals for your future product.
IV. **Analysis**. Ask why you and the competition have not yet achieved the proposed targets/goals.
V.   **Proposed Countermeasures**. Define your proposal to reach the future state. Consider that the product to be developed might be very complex, and encompass the creation or modification of the actual value chain.
VI. **Plan**. Based on the key metrics, you can develop the next actions, the timeline and the responsibilities. We recommend the use of milestone plans, setting the timeline and responsibilities to the next future target conditions.
VII. **Follow up**. Describe the planned actions to follow up the A3 execution and to trigger the beginning of the next planning cycle.

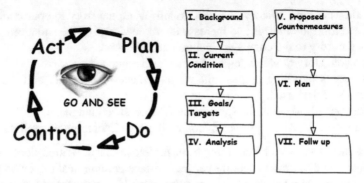

Fig. 8.5 The problem-solving A3 and the PDCA

 8.5 A Practical View

In this chapter, we discussed the organizational structure and responsibilities of Toyota and presented some knowledge management tools and techniques. Remember that more important than the concepts themselves is the reason and value behind them.

The presented structure (Fig. 8.3) embeds some important aspects that guarantee effective knowledge management and which can give you insights of what can be done in your own organization:

- Having common objectives which are the basis for everyone's evaluation are what fasten and align all the pieces in the structure.
- The weak matrix also supports functional (vertical) learning since all the lessons learned can be timely exchanged among the functional program teams' integrands.
- By creating the module development team, cross-functional (horizontal) learning is also supported.
- Special attention and respect is given to the supplier, where both the supplier and the company benefit from creating and from the created knowledge.

Considering the KM tools and techniques presented, we recommend its use during the LPD as shown in Fig. 8.6, where:

- The partitioning of the development scope into smaller cycles also increases the learning pace (rapid learning cycles) while it reduces the rework cycles in the development.
- On a daily basis, the techniques of 5 Why, cross checking, problem solving at the source, and *ijiwaru* are kept in mind and used whenever they are necessary. Their use is not considered a waste of time, but an opportunity to learn and to not repeat the same mistakes again.

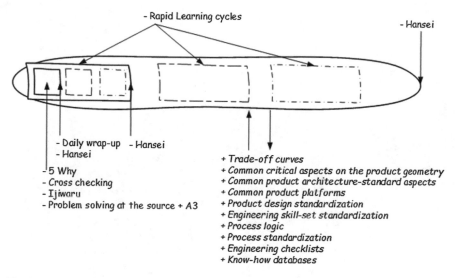

Fig. 8.6 A practical view

- At the end (or at the beginning) of each day's work, the team discusses its last achievements and issues and sets the plan ahead.
- The practice of *hansei* is widely used, not only when the development is finished but also at the end of each cycle and every day.
- All the previously documented knowledge supports the related activities while they are themselves updated in order to include the last lessons learned.
- The process knowledge techniques support the PDP through its execution.

References

1. Ward A (2007) Lean product and process development. The Lean enterprise Institute, Cambridge, MA
2. Morgan JM, Liker JK (2006) The Toyota product development system. Productivity Press, New York
3. Rother M (2010) Toyota kata: managing people for improvement, adaptiveness and superior results. McGraw-Hill, New York
4. Gulati RK, Eppinger SD (1996) The coupling of product architecture and organizational structure decisions. [S.l.]. MIT Sloan School of Management Working Paper, No 3906, 28 May 1996
5. Cusumano M, Nobeoka K (1998) Thinking beyond lean. The Free Press, New York
6. Shook J (2008) Managing to learn: using the A3 management process. Lean Enterprise Institute, Cambridge

Part IV
The Tire

Part IV discusses the Lean Product Design and Development Process itself and its related tools and techniques (Fig. 1). Chapter 9 shows a general view of the LPDDP and introduces the following chapters. Chapters 10, 11, 12 and 13, discuss each of the LPDDP phases and groups of activities, and present some tools and techniques to be used while performing the process.

The Value Function Deployment (VFD) and the Product Development Visual Management Board (PDVMB) are two techniques presented in Chap. 9 that support our approach to the Lean Product Development Process. Our experience shows that these techniques are very useful for supporting the transitioning from a tradition PDDP to a LPDDP.

Fig. 1 The lean wheel's tire elements

Chapter 9
The Lean Product Development Process

This chapter describes the product development phases and activities, which are detailed in Chaps. 10–13. We consider four phases in the PDP: (1) the portfolio phase, which produces a general vision of the product, both aligned to the value pulled by the market/customers and consistent with the company's strategy and capacity; (2) the study phase, which includes the identification of the value pulled by both by external and internal stakeholders, the value proposition activities that outline the chief engineer's vision of the new product, and the value delivery planning for the next phases; (3) the execution phase, including the design, development, production/ramp up of the products and/or services that deliver the pulled value; and (4) the use phase when the resulting product/process is followed-up until its discontinuation. The Value Function Deployment (VFD) technique and the Product Development Visual Management Boards (PDVMB), which are also presented in this chapter, support the Lean Product Development Process execution.

9.1 Introduction

Womack and Jones [1] noted that the ideal process of designing a product should function congruently with single-piece flow in manufacturing. It suggests that this process should represent a continuous flow of value creation, from conception to production, without stops due to paperwork and no returns for error correction.

From their study of the Toyota Product Development System, Morgan and Liker [2] identified two main phases in the Toyota's lean product development process: (1) the study phase, *kentou*, and (2) the execution phase.

During *kentou* the PD teams can anticipate, study, and resolve problems, completing such tasks as fundamental design decisions, identifying failure modes, designing in countermeasures, and setting cross-functional objectives. *Kentou* results in far fewer engineering changes and creates process flow by allowing companies to focus on downstream task execution. It also provides a formal

© Springer International Publishing AG 2017
M.V.P. Pessôa and L.G. Trabasso, *The Lean Product Design and Development Journey*, DOI 10.1007/978-3-319-46792-4_9

structure for cross-functional teams to "design in" solutions, which is far less expensive than solving problems or "fixing" designs later in the process. During the study phase, the product is conceived, a performance envelope is defined, and the solution space is explored in order to find a balanced (value/risk) design.

Once *kentou* is complete and the development strategy is set, the execution phase may begin. By the time it reaches this point, the LPDO has made a full commitment to the product, and has begun to invest significant sums of money in tooling and in its suppliers. Because of this investment, it is financially critical to have a high-velocity PD process with radically shortened lead times, by focusing on precise execution and smoothing product-to-market delivery [2]. The company's goal from this point forward is to optimize capital investment, match quick cycle-supporting or embedded technology lead times, make decisions closer to the customer and other relevant stakeholders, and react quickly to changes in the competitive environment. Creating flow by synchronizing product development activities is one of the most powerful ways to increase speed.

Rather than describing how Toyota works, our objective here is to help companies implement lean PD systems themselves. While the Value Function Deployment (VFD) technique, also described in this chapter, is the backbone of our implementation model, you can use other ways to achieve similar results provided you keep the same philosophy. Therefore, the book's proposal is to focus on the concepts of continuous improvement, value delivery, and waste reduction, as presented in Part II, while keeping in mind the cultural, organizational, and knowledge management aspects, as discussed in Part III.

9.2 The Process and Its Phases

The PDP model we use here aims to:

1. support the practical application of the concepts previously described in Parts II and III of this book; and
2. fit the VFD and the PD Visual Management Board (PDVMB), which are further described in this chapter.

Even though most tools and techniques can be used in the lean way experience shows that is very difficult for a person used to applying tools and techniques with the mindset bounded by a certain paradigm to do that in a different way. Unconsciously he or she turns back into the previous way. This is the reason we proposed the Value Function Deployment (VFD) technique [3, 4]. The VFD acts as a backbone of the Product Development Process, always reminding the practitioner about the lean directives while he/she can apply the tools and techniques he is accustomed to.

In the same way, the presented PD Visual Management Board (PDVMB) is a sample of simple *obeya*, which provides visual management. Our experience also shows that people struggle to initially define what is important to be included in

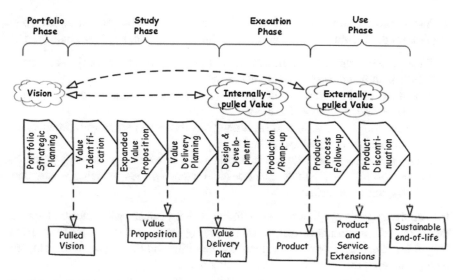

Fig. 9.1   Our PDP model

the *obeya*, and how to make the management by using it. The PDVMB is a start-ing point to defining your own *obeya*, and include what we believe is the minimum information to manage a LPDDP.

We divide the product lifecycle in four phases (Fig. 9.1), where the output from each phase is both aligned with (1) the value pulled from the final user/customer and other external stakeholders, and (2) the value pulled from the subsequent phases into the PDP (internal stakeholders). The phases are further detailed into groups of activities.

The phases are briefly described in sequence. The study and execution phases' activity groups are further detailed in Chaps. 10–13.

9.2.1 Portfolio Phase—Portfolio Management Activities

This phase includes all the portfolio management activities and ends by delivering a "product vision" which presents a general description of the expected develop-ment results and their market impact plus any constrains and assumptions initially bounding the product development conceptual work.

The result from this phase is a general vision of the product, both aligned to the value pulled by the market/customers and consistent with the company's strategy and capacity.

Good portfolio management is a key success factor to the LPDO. Portfolio management is about resource allocation (how your business spends its capital and human resources) and project selection (ensuring that you have a steady stream of big new product winners). Therefore, portfolio management has four goals [5, 6]:

1. Guarantee the strategic alignment where the final portfolio of projects is strategically sound and truly reflects the business's strategy.
2. Maximize of the return of the investment (both in terms of the company's objectives and, of course, the money).
3. Balance (long/short term and high/low risk) the development programs in the various markets the business is in.
4. Create a development cadence that balances value delivery through products/markets and the company's resources and capacity, thus reducing waste, unevenness, and overburden.

In fact, these goals act like valves defining which projects will enter and stay at the product development funnel, while regulating the flow of development projects (Fig. 9.2). These development projects can be either the development of complete new products or the improvement of existing ones.

The strategic alignment is guaranteed by taking into account that all development projects respond to both value pulled by the customers/market and the value pulled by the shareholders. In order to do that, the company needs to have a clear vision about itself, its products, and the related technologies it wants to master. Business, product, and technology governance play an important role at this

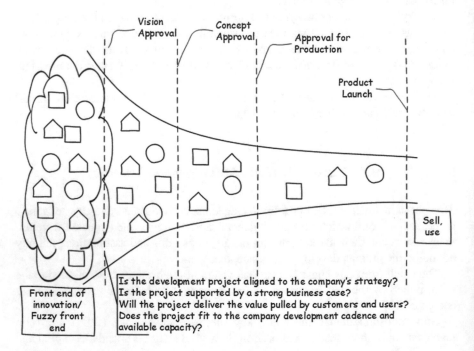

Fig. 9.2 The portfolio management goals through the funnel

moment since they can guarantee not only the individual development projects' alignment, but that the projects from the portfolio have a positive impact on each other.

As a rule of thumb, the LPDO should avoid developing new technology in individual development program critical paths. Therefore, technological innovation is strategically focused, often in response to a request from a chief engineer, and aims to create off-the-shelf proven technology. In the case technology development cannot be avoided, the incorporating of this new technology should be considered as one of the SBCE alternatives, as presented in sequence and detailed in Chap. 11.

Considering the PD funnel, the LPDO only triggers the concept development (normally done by the chief engineer and his/her staff) after a strong business case (a "value case") is achieved, and respects the cadence discipline (i.e. when the company, for a certain product, releases periodical updated product versions, like cars, cell phones, etc.). Not respecting the company's development capacity might lead to waste, unevenness, and overburden through the development portfolio. These ripple effects are one of the main causes of firefighting through and across projects.

9.2.2   Study Phase—Value Identification Activities

After receiving the Product Vision, the chief engineer or its equivalent starts the study phase's value identification activities which aim to provide deep understanding of the true value to be incorporated into the product (and/or service). All the related stakeholders through the value chain, both internal and external, must be considered and the value they pull understood. As a consequence, all the stakeholders from the use and execution phases should be listened to.

Different stakeholders have different importance; also, any pulled value item is associated with some risk (business, market, technical, etc.). Sometimes the identified value challenges the vision-related constraints, so if a trade-off solution is not achievable, the vision must be challenged (negotiated).

The objective of the value identification activities is to create a structured and unambiguous value items set, rooted in the stakeholders' pulled value, and which serves as reference for all the development team, therefore guiding all the development program activities.

This is the moment during the study phase when the value pulled by all key stakeholders is consolidated. In this process, explicit and implicit agreements are made in order to balance all stakeholders' needs, resolve conflicts, and include tangible and intangible values (protecting the environment, meeting the technical specifications, meeting the shareholders' expectations, providing an environment of rewarding work, etc.), or anything that has been forgotten.

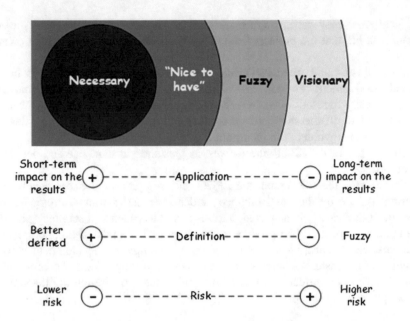

Fig. 9.3 Requirements classification

The value items set formalize the program goals, defines relationships between stakeholders, and sets the cost and time constrains. It creates the program "flight envelope" to deliver the desired value.

Even though the value, as perceived by the customer, is the primary pull force to the whole development, only by considering the value stream based on the needs from all key internal and external stakeholders can you ensure that all the people, groups, and businesses which can impact decisively in the development project will contribute with work and resources to ensure the project's success. Often, the efforts to make the value visible for the various stakeholders require the inclusion of additional development. These activities are though necessary, since anyone who does not perceive receiving any value will tend to stop contributing.

Each development project carries some risks. In fact, since several and sometimes conflicting pulled value might arise, the LPDO must discern the real value (the ones that will trigger the buying decision) from: (1) "nice to have" features/characteristics which consume resources and increase the development risk; (2) fuzzy requirements which still are not completely clear, maybe even in the customer's mind; and (3) visionary requirements which may require great architectural or technological changes and the related risks (Fig. 9.3). Any requirement out of the "real value list" could postpone the development and negatively affect the planned cadence. Considering or postponing (in the case where a new version of the product is released from time to time, i.e., annually) the other requirements should be balanced against the associated risks. In any case, the strategy chosen should allow dropping them at any time and with minimum impact on the development flow.

9.2.3 Study Phase—Value Proposition Activities

The Value Proposition outlines the chief engineer's vision of the new product; communicates customer-defined value and product-level performance objectives, and aligns the product-level performance goals of the entire program team; in summary, it communicates all the pulled value in a simple, unambiguous, and final written document [2].

The product development project aims to deliver the value proposition which might range from a particular product to a completely new or modified value chain.

Often during the execution of a development project, the development team might face conflicting pulled value issues. If the development project lacks a clear value proposition where priorities are set, the team might make decisions by having only partial knowledge.

In order to finalize the value proposition, the next challenge is to define which functional architecture is the preferable choice to deliver the product/service.

In most cases, PD is an open-ended problem, therefore accepting multiple possible solutions. As a consequence, each of the product's functions can be implemented in different ways. These alternatives, though, carry intrinsic risk, so the LPDO must carefully chose the path to follow, since this can be the difference between success and a huge failure.

In order to reduce the chances of iterative loop-backs or plan modification and resource changes, which might create ripple effects in the whole company's PD portfolio, the LPDO uses SBCE (see Chap. 6). As a consequence, the value proposition might include different product's subsystems alternatives, where the set have very low chances of causing rework loop-back due to failures of all alternatives. SBCE explores the solution space, supports the no-compromise attitude, allows emergent solution (combining) and creates knowledge.

Experience shows that the cost of applying SBCE is equivalent to applying the point-based approach, considering the average needed rework cycles. The great difference among them is that SBCE greatly reduces the risk of overtime, while generates more knowledge by understanding the several design alternatives of the product.

Although equivalent, SBCE requires, though, more resources to carry out simultaneously the different product design alternatives. These resources might not be available in all companies, therefore the need of prioritizing in which product's parts/modules/subsystems to apply the SBCE. We consider that the product's parts/modules/subsystems which deliver more value and/or are more risky as critical to applying SBCE (see Chap. 11 for details about the prioritizing strategy).

9.2.4 Study Phase—Value Delivery Planning Activities

During the value delivery planning activities, all the teams that will work on the project are defined and a set of pull events is determined. By having the teams and pull events, it is possible to create a plan that embeds real concurrent engineering and flow.

As we mentioned before, in the traditional PDP, the development plan is followed until it fails (point-based), for whatever reason, and then follows a series of iterative loop-backs, or plan modification and resource changes. As a consequence, the results from the work performed during the execution phase are pushed through the activities. A systemic view of the solution (and often only part of it) is only achieved in phase gates. These gates, besides damming information, often lead to unnecessary delays and inventories.

The lean principles state that no process along the value flow should produce an item, part, service, or information without direct request from the afterward processes. By pushing results through the PDP, the company is just accumulating a stock of information and items that no one wants yet and that might become obsolete before being used.

The best way to understand the logic and the challenge of pull production is to start with a real customer expressing a demand for an actual product and walk the other way, going through all the steps required to bring the product to the customer.

This promotes high flexibility, allowing all the activities along the process to produce exactly what the customer (either internal or external) wants and when he wants it. Moreover, the reduction in response time for fulfilling the consumer needs speeds up the return on investment and reduces inventory even in a complex production flow.

Applying the "real" pull system concept into product development is a challenge. Each development project is unique; therefore, there is no fully predetermined way of how to build a product. Subsequent processes cannot pull definite information from its predecessors, since they are neither aware of the outcome of the work they will perform, nor of the final product with all its specifications.

It is possible, however, to get a good feel for what to expect, since the activities follow a logical sequence and the history from previous similar projects and information gives a good idea of the necessary inputs and outputs to be generated.

In the case of a development project, the important thing is to let the customer pull the value of the performing team. To make this possible, the development activities must be connected in a simple way and help eliminate waste from them.

As a consequence, a pulled value delivery planning ensures progress and project quality. Instead of phase gates, which dam lots information and stop the flow, pull events based on tangible results such as models, prototype ready systems, etc. allow the flow [3, 4].

The pull events relate directly to the value items, i.e., the scope of an event is associated with valuable items and their effectiveness measures. Unlike phase gates, pull events are part of the development value stream, and cannot be eliminated.

As a result, pull events have four roles, namely they (1) determine that there has been progress in the effective delivery of value; (2) ensure that information on the project will be pulled, not pushed; (3) allow the combination and the strengthening of alternatives during the SBCE; and (4) are learning moments as they allow reflection (*hansei*) about the progress of the work and the results obtained through adopted strategy.

9.2.5 Execution Phase—Design and Development Activities

At this phase, all the module development teams will produce their deliverables in a fast and synchronized way according to the sequence of defined pull events.

The LPDO maximizes the return of the investment by guaranteeing that the product to be developed has been pulled by the customer and that the value chain is aligned both to the goal and within itself. As a consequence, after starting the execution, the LPDO uses decision analysis (e.g., cost/benefit) to find the best alternatives to keep going rather than for deciding whether it should continue or stop the project. Design strategies (i.e., DFX and DTX) aligned to the pulled value set also multiply the impact from the development effort and expedite the return of the investment.

Pull events foster concurrent engineering, are opportunities to reveal quality problems, and support knowledge creation. In this context, planning is decentralized, allowing different groups to realize their own plans to achieve the pull events. For example, narrowing the sets of points in SBCE are pull events.

As a consequence of continuous improvement, the PDP can be seen as a spiraling and iterative process through the PD funnel, each cycle corresponding to one PDCA round (Fig. 9.4), where the "C" corresponds to a particular pull event.

9.2.6 Execution Phase—Production/Ramp-up Activities

The activities on this phase will drastically change according to the kind of product and the related production expected rate. One-of-a-kind products, for instance, can be even the final prototype from the development phase.

During ramp-up the product production and service delivery begin. Energy supplied, manpower deployed or quantities produced are gradually increased. At this moment the production process is proven, and there might be change request to adapt either the product or the production process to support full power production.

Once these initial issues have been solved, production is adjusted to fulfil the market demand.

Fig. 9.4 Product
development as a PDCA

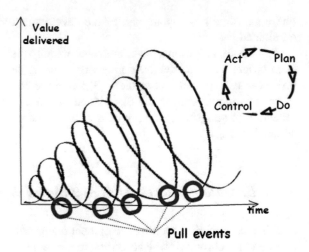

At this phase the tools and techniques from the Toyota Production system are fully applied, and the development system tools and techniques are only necessary when a product/process change is requested.

9.2.7 Use Phase—Product/Process Follow-up and Product Process Discontinuation Activities

Even though this is the last phase to actually happen, all the PDS is based on it. The initial understanding of the use phase triggers the portfolio phase in order to consider this perceived need a candidate for a product development project. Even after a selection is made during the portfolio phase, the understanding of the use phase is further explored during the study phase in order to guarantee that the product to be delivered will match the pulled value. This phase includes the "Product/Process Follow-up Activities," and the "Product Discontinuation Activities," which comprise the product use, training, maintenance, evolution, and discontinuation.

9.3 The Value Function Deployment—VFD

The Value Function Deployment (VFD) [3, 4] technique described in this section applies the lean principles based on value creation and waste reduction to derive a project activity network that entails a sequenced set of confirmation events. These events pull only the necessary and sufficient information and materials from the product development team.

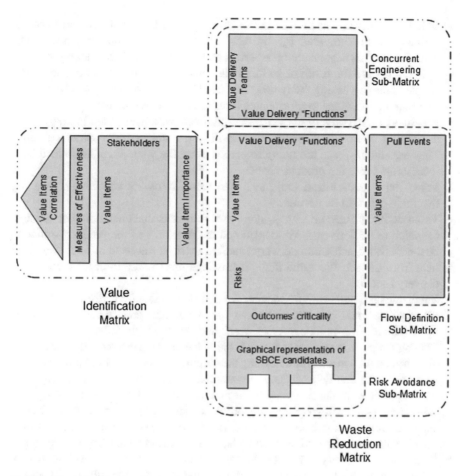

Fig. 9.5 The value function deployment matrices

The VFD is an adaptation of the Quality Function Deployment (QFD) technique and is composed of two interconnected matrices, the value identification matrix and the waste reduction matrix (Fig. 9.5). The former captures, prioritizes, and shows the correlation between all the value items expected by the project's stakeholders. The latter deploys the value items to the value delivery functions, calculates their criticality (rework avoidance sub-matrix), correlates the functions to the teams responsible to implement them (concurrent engineering sub-matrix), and defines the events that will pull this value from the teams (flow definition sub-matrix).

The VFD matrices' core elements are defined in sequence:

• **Stakeholders** are individuals or organizations that are actively involved during the development or whose interests may be affected by its execution or outcome.

- **Value** for a given stakeholder is the complete and balanced perception of the various benefits provided by the results from the development process. The value is stated in the stakeholders' terms and might not be free of ambiguity.
- **Value items** are the result of splitting the value into more specific and measurable elements (by asking "why you need that?" or "what do you mean by that?"). They can be functions, performance, level of acceptable risk, etc.
- **Measures of effectiveness** (MoE) are reference parameters used to analyze the conformity of the PDP results in relation to the stakeholders' expected value. They explain how you are going to perceive that the value item has indeed been incorporated into the product/service.
- **Value items correlation** indicate if two value items are conflicting, meaning that trade-offs will be needed.
- **Value delivery functions** are system level functions that encompass or relate to the value to be delivered. We considered the functions of the product/service to be developed. Each value delivery function must be traced to at least one value item from the set; the value items themselves must relate to at least one value delivery function.
- **Value delivery teams** are responsible for delivering value by performing the value delivery functions. We divided the teams in two groups: those which deliver value by developing the product/service itself, and those which deliver value by performing supporting processes through the value chain. Therefore, the part of the organization responsible for designing the specific subsystems of the product populates the teams that deliver value via *product*. Similarly, the part of the organization responsible for the designing of the processes that deliver the value through the value chain (such as marketing, supporting services, etc.) are the teams delivering value via *processes*; indeed, these processes are paramount for the stakeholders to perceive that they obtained the total pulled value of the obtained project's benefits.
- **Outcomes criticality** refers to the amount of value and the level of risk to deliver this value by each value delivery function. As a result, the functions which deliver more value and/or are at more risk are the most critical ones.
- **Pull events** typically are tied to physical evidence of progress (presentations of models, prototypes, initial production, etc.). We recommend using: (1) integration events that create "boundary objects" as built engineering projects, mock-ups, prototypes, etc.; (2) successful endings of checks and validations, which are moments of reducing uncertainty and risk in the program. The pull events set creates a "ladder," where each step gets closer to the development success.

Considering the presented PD lifecycle, Table 9.1 shows how the VFD is applied during its phases.

Table 9.1 VFD matrices and the PD lifecycle

VFD matrices	PD study phase—activities groups
Value identification matrix	Value identification
Rework avoidance sub-matrix	Value proposition
Concurrent engineering sub-matrix	Value delivery planning
Flow definition sub-matrix	Value delivery planning

9.3.1 Value Identification Matrix

The VFD is centered on the value pulled by the stakeholders. The Value identification matrix provides a straightforward visualization of all the value items pulled by the stakeholders, how each value item can be measured during the development, how the value items correlate to each other, and their relative importance for the development. The value identification and grouping is divided into five steps (Fig. 9.6):

1.1 **Identify the stakeholders**: All the stakeholders, both external and internal, must be considered. External stakeholders are those related to the use phase, while the internal are those related to the execution phase. Failing to recognize the external stakeholders may jeopardize the products' market success. Failing to recognize the internal stakeholders may compromise the concurrent engineering and smooth product development, production, and logistic flows.

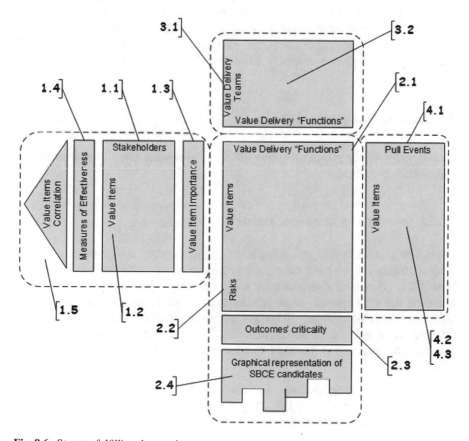

Fig. 9.6 Steps to fulfilling the matrices

1.2 **Analyze the value items**: This step includes understanding the stakeholders' needs and breaking them down into value items. The value items remove the ambiguity from the value set so the items can be addressed by the teams' deliverables and the progress on effectively delivering them can be measured. For example, a need presented as "be safe" can be broken down into items corresponding to the homologation tests defined by the product's regulatory agency.

1.3 **Prioritize the value items**: Each considered stakeholder has particular needs and thus rates the importance of the value items differently. The value items prioritization takes into account the combination of these ratings.

1.4 **Define measures of effectiveness (MoE)**: At least one measure of effectiveness must be defined for each value item. These measures allow the verification and validation that the value items were effectively incorporated into the project's results.

1.5 **Identify conflicting value items**: Conflicting value items are items that cannot be optimally delivered simultaneously (like having a car with high speed and low fuel consumption at the same time) if using the current company knowledge and capacity. The conflicting value items direct the creation of trade-off curves that, besides aiding the development team, are part of the company's knowledge assets. By challenging and improving the trade-off curves, a company becomes more competitive.

9.3.2 Waste Reduction Matrix

The objective of the waste reduction matrix is to support ways to reduce rework and guarantee the flow.

9.3.2.1 Rework Avoidance Sub-Matrix

The development of multiple alternatives prevents the early abandonment of promising solutions while giving room to the coexistence of preconceived alternatives. The SBCE helps guarantee the flow while reducing rework cycles: if one alternative on the set is proven to be inadequate, the others can still be used and no additional work is necessary. This process determines the most critical product functions or organizational value chain functions that will be developed through a set of alternatives, and is divided into three steps (Fig. 9.6 maps these steps on the VFD matrices):

2.1 **Define the value delivery functions**: This step determines the product's functions which deliver the complete value items set. Each function must contribute to delivering at least one value item and vice versa.

2.2 **Address risk response**: Identify the risks related to successfully delivering the development project results. The risks might relate to either incorporating the value into the functions themselves or issues that might arise during the development project management.

2.3 **Calculate the criticality of each value delivery function**: The functions' criticality is directly proportional to: (1) the amount and importance of value to be incorporated in these functions; and (2) the perceived risk to successfully deliver the expected value subset. The more valuable and the more risky, the more critical the functions are.

2.4 **Define the priority to parallel development**: The functions to be developed through a set of alternatives will be chosen by considering the restrictions imposed on the development project and the previously calculated criticality. The definition of the number of alternatives and the characteristics of each of the alternative will take place during the execution phase.

9.3.2.2 Concurrent Engineering Sub-Matrix

The strategy of using the functional architecture as the basis for determining the development team structure has great advantages for the application of SBCE. In this case, one team must determine the various alternatives, unlike functional organizations where this responsibility can be distributed among various groups, hindering the SBCE control.

The relationship between value functions and value delivery teams determines the need for concurrent engineering. This occurs because the effective delivery of a particular value item can depend on incorporating the results into different value delivery functions which are the responsibility of different teams.

This sub-matrix is divided into two steps (Fig. 9.6):

3.1 **Identify the value delivery teams:** This step determines which teams are responsible for the delivery of each function. These teams are either related to the product subsystems themselves or to organizational processes (such as marketing, quality, production, etc.).

3.2 **Define the contributing roles of each value delivery team**: This step maps the role of each team on delivering a particular function. After completely filled, this sub-matrix works like a Role & Responsibility Chart (RACI).

9.3.2.3 Flow Definition Sub-Matrix

No process along the value flow should produce an item, part, service or information without direct request from the afterward processes. The pull events are the backbone of the value flow and are important moments to knowledge capture; by pulling the value delivery, they allow the planning to reach execution. Every pull event is associated with physical progress evidences (i.e., models, prototypes, start of production, etc.). The pull event determination process is divided into three steps (Fig. 9.6):

4.1 **Define preliminary pull events:** To define a sequence of preliminary pull events, the development team can use the enterprise's standard process (if there is one), reuse historical information from previous projects, or consider best practices from the industry.

4.2 **Relate the pull events to the value items and risks**: A pull event scope is defined by the set of value items and risks it will check and how they will be checked (i.e. analysis, subsystem tests, integrated tests, etc.). A pull event must be related to at least one value item and/or risk, and each value item/risk must be checked by at least one pull event.

4.3 **Refine the pull event set**: The preliminary pull event set is refined until it meets the following criteria: (1) it must be capable of verifying the progress on the effective value incorporation and delivering during the project execution; (2) it must represent the value flow in order to guarantee the information pull, and not push; and (3) it must show the elimination of the risks that led to the development of multiple alternatives, allowing the combination and the reduction of the number of alternatives during the SBCE.

9.3.3 Systems Engineering and the VFD

During the definition and decomposition of the system to be developed system engineering design activities detail the system using a top-down approach, from conceptual design to detail design. The previous VFD description was made at the conceptual level, once it relates the pulled value to the value delivery functions. In order to guarantee value traceability and consider the SBCE risk reduction capabilities through the design phases, the VFD can and should be used at all design stages:

- Conceptual design (system-level): this is what we have already done in the study phase, when we checked which of the systems' functions were best candidates to SBCE, and looked for possible alternatives for supporting the subsystems that would perform these functions.
- Preliminary/layout design (subsystem-level): in the same way, the teams in charge of each subsystem can check which of its constituent modules are more critical. At this moment, a different VFD is built for each of the subsystem's alternatives. Depending on the system complexity, this breakdown has to be done in several steps, once the modules might be composed by submodules.
- Detail design (module level): Detail design goes until you reach parts definition. SBCE can also be applied to the most critical parts from each module, where alternatives might include chosen different parts, materials, or suppliers.

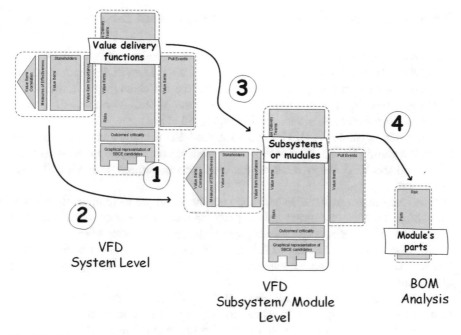

Fig. 9.7 Using the VFD during the design of the product/service

Figure 9.7 emphasizes some of the critical aspects when detailing the VFD into further detail levels:

1. Critical functions, which were chosen to SBCE, will lead to different subsystem alternatives; therefore one subsystem-level VFD has to be built for each of these alternatives.
2. Only the value that is related to the further levels of detail is carried out. Note that some internally pulled value items might be added, which is the case of including DFX directives. Externally pulled value items, though, can only be added at the system-level VFD, once they potentially impact the whole product/service.
3. The value delivery functions should be grouped into subsystems, which can be physical products or services, and these subsystems are further detailed during the design.
4. When analyzing the parts (Bill of Materials – BOM) there is no need to build a complete VFD, once the analysis is centered in the risk (different part numbers, materials and suppliers).

9.4 Product Development Visual Management Boards

As presented in Chap. 7, the *obeya* (big room) is a good example of visual communication. In order to support putting in practice the Lean PDP depicted herein, we developed *obeya* models to be used during the activities from the Study and Execution phases. These Product Development Visual Management Boards (PDVMB) function as continuously developing A3 charts.

The study phase PDVMB (Fig. 9.8) has the VFD filling as its focal point, the execution phase PDVMB (Fig. 9.9) has the product under development as its focal point, and the VFD keeps track of the development project progress and value alignment.

The study phase of the PDVMB supports the value proposition creation; communicates stakeholder-defined value, product-level performance objectives; and aligns the product-level performance goals of the entire program team.

The execution phase of the PDVMB keeps track of the product evolution during its design and developments and supports the concurrent engineering and change management.

Both the study and execution phase of the PDVMB include quality, time, and cost indicators. Quality is represented in the "compare current product/competitors/substitutes/new product" field, by showing the planned versus designed/developed product value delivery capacity. Time and cost can be tracked by creating an "S-curve" from the milestone chart.

The detailed PDVMB filling is explained in Chaps. 10–13.

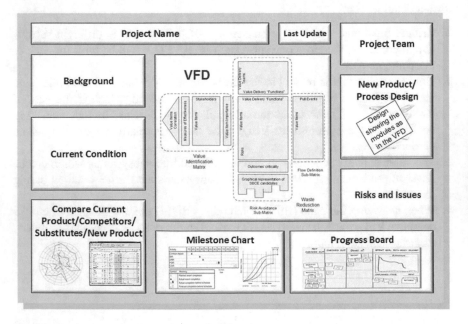

Fig. 9.8 Study phase visual management board

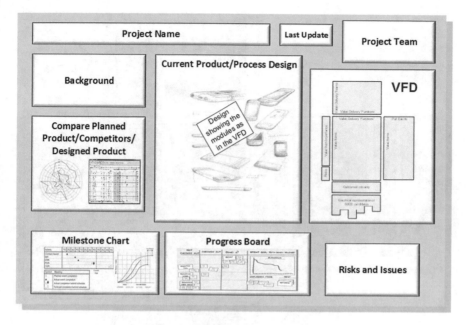

Fig. 9.9 Execution phase visual management board

 ⇨ **9.5 A Practical View**

At first sight, the VFD looks a bit complex and hard to fill. You must remember, though, that product development is itself a complex task and that the VFD filling is gradual. Consider, for instance, a project management plan with all the related process areas (time, cost, quality, risk, procurement, etc.). Looking at the VFD is like looking at most of those areas at the same time and they are integrated. Indeed, the VFD visually presents and supports answering some key development questions (Fig. 9.10), as presented in Chaps. 10–13:

1. What is the comparative importance of the value items among themselves, considering their relevance to the considered stakeholder set?
2. What value items conflict with each other, thus bringing the need of trade-offs?
3. How am I sure that the functional architecture (product and value chain) is capable of delivering all the pulled value?
4. To what functions should I give more attention once they are more critical (deliver more value and/or carry out more risk)?
5. How can I determine the need of concurrent engineering and who has to work together and when?
6. How can I define a balanced development execution strategy which covers the complete scope and considers all the risks?

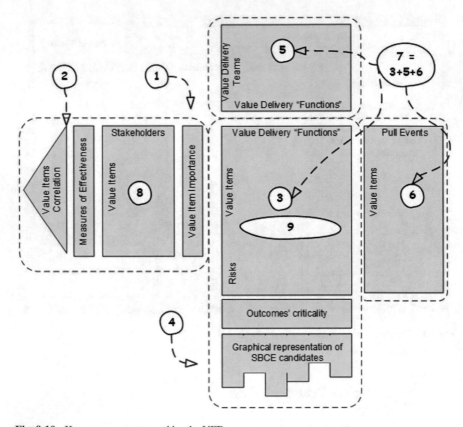

Fig. 9.10 Key answers supported by the VFD

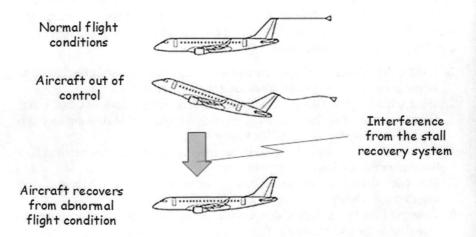

Fig. 9.11 The stall recovery system in action

7. How can I determine the "waste less" set of activities to be performed by all the teams?
8. How can I identify the need of DFX?
9. How can I identify the presence of integrative variables and the need of DTX?

In Chaps. 10–13 we use a product development example that illustrates the PDP being supported both by the VFD and the PDVMB. The data was collected from a finished and successful project which produced a stall recovery system to be used during flight tests and which had the objective of recovering the aircraft to normal flight conditions (Fig. 9.11) in case the pilots lose control of the aircraft while performing flight tests of a prototype aircraft.

References

1. Womack JP, Jones DT (2003) Lean thinking. Free Press, New York
2. Morgan JM, Liker JK (2006) The Toyota product development system. Productivity Press, New York
3. Pessôa MVP (2006) Proposta de um método para planejamento de desenvolvimento enxuto de produtos de engenharia (Doctorate Thesis) Instituto Tecnológico de Aeronáutica: São José dos Campos
4. Pessôa MVP, Loureiro G, Alves JM (2006) A value creation planning method to complex engineering product development. In: Proceedings of 13th ISPE international conference on concurrent engineering, Antibes. Leading the web in concurrent engineering, vol 143. IOS Press, Amsterdan, pp 871–881
5. Cooper GC, Edgett SJ, Kleinschmidt EJ (2001) Portfolio management for new products, 2nd edn. Basic Books, New York
6. Project Management Institute, PMI (2013) The standard for portfolio management, 3rd edn. Project Management Institute, Newton Square

Chapter 10
Study Phase—Identification Activities

A product development project normally has its genesis after a PD project is included in the company's development portfolio, and the product vision is given to the chief engineer. This pulled vision puts the PD wheel in motion, triggering the identification activities (Fig. 10.1), which aim to deliver a comprehensive description of the development project scope, by gathering the value pulled by all the involved stakeholders, both external and internal to the development. This chapter uses the stall recovery system project example to present a stepwise execution of this phase's activities, where special emphasis is given to eliciting and prioritizing the value pulled by the external and internal stakeholders.

10.1 Introduction

In many traditional companies, once given the product vision, the development team will (1) assume it like law; (2) try to produce the concept internally, mixing the experience and expertise from the team members; or (3) use some market and business data to support their decisions. The lean way assumes the vision as a hypothesis to be confirmed or refuted. The team must go and see (*Genchi Genbutsu*).

Even though we use the PDVMB and the VFD to support the value identification activities, you can use other tools and techniques, provided you keep the lean philosophy. But we strongly recommend you to do it as we suggest here, at least on some of your PD projects, thus gaining confidence to try other approaches and keeping the lean philosophy.

By following the PDVMB filling steps, the development team is guided through the lean journey. It acts as a direction giver, providing process discipline, fostering communication, and facilitating management.

© Springer International Publishing AG 2017
M.V.P. Pessôa and L.G. Trabasso, *The Lean Product Design and Development Journey*, DOI 10.1007/978-3-319-46792-4_10

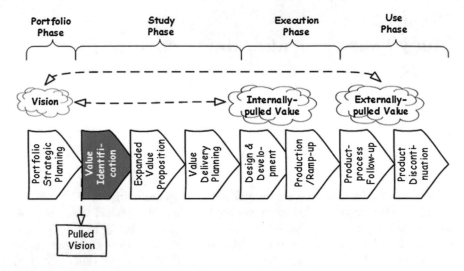

Fig. 10.1   Value identification activities position in the PDP

10.2 The Board Guides the Team Through the Lean Journey

The PDVMB and VFD filling sequence described below (Fig. 10.2) will guide you during the value identification activities. Note that at this moment only the Value Identification Matrix from the VFD is filled, and the new product/process design is not filled.

10.2.1 The Background Is the Basis for Everything

The contents of this field are normally given to the chief engineer when he/she receives the request to lead the project. It can arrive in different formats: a direct request, one objective from the company strategy, a customer's request, a project charter, a signed contract, etc. Regardless of the case, it can be always summarized into a couple of unambiguous statements describing:

- What is the product vision?
- What versions/models are available to customer?
- What options/groups of options would be made available to the customer as add-on modules or services?
- Is this product the first of a future product line where we need to create a platform upon which to build other offerings in future projects?

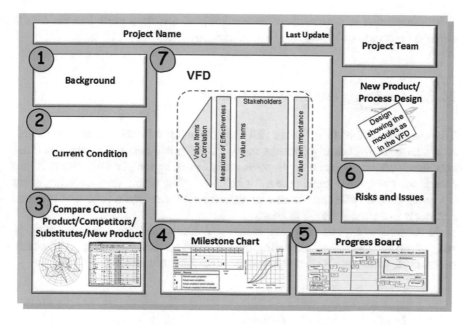

Fig. 10.2   PDVMB filling sequence

- What components/subsystems of the product must remain fixed from the prior product? (i.e. must be reused or not changed due to part commonality with other models, safety, cost, packing, etc.)
- What are the current product and its history in the marketplace? If there is no current product, what is the related previous experience that the company have?
- Why are we doing a new product/product line? What demand/value do we perceive as pulling the new product development?

In good systems and requirements engineering, the product vision corresponds to the Business Requirements Level [1, 2]. Therefore, it should describe the sponsor's point-of-view, and define the objective of the product development project (goal) and the measurable business benefits for doing the project:

> **The purpose of the** [project name] **is to** [project goal—that is, what is the team should implement or deliver] **so that** [measurable business benefit(s)—the sponsor's goal].

For instance, this book's background was described as:

> **The purpose of the development project entitled** "THE LEAN PRODUCT DESIGN AND DEVELOPMENT JOURNEY: A PRACTICAL VIEW" **is to** develop a book, which is going to be available both in printed online versions, and **which aims to** fill the literature gap of addressing a method to support PD practitioners while changing their current PDP into Lean PDP.

> In order to be really practical, we will base our arguments on our experience both on practicing and teaching PD, and will take advantage of our previously published academic work, which was already peer reviewed

In the case of a process development project, this field must include details about the process AS-IS, and why it needs improvement.

10.2.2 Analyzing the Current Condition

The contents of this field (Table 10.1) are normally available at the chief engineer's request. It basically contains market intelligence data which supported the new development project attractiveness analysis during the portfolio phase. If this data is not available, special care must be taken while identifying the value, since the development vision may not be supported by relevant and reliable market data, and might not sustain a business case.

In the case of a process development project, this field must include the mapping of the process AS-IS, including the identified wastes, its impact (particularly on the cycle time), the bottlenecks and the process restriction.

10.2.3 I, Myself, and the Others

On the comparative board, the company's actual product (if any), the competitors, the substitutes, and the planned new product are compared according to the customers' and the final users' pulled value.

This board will be updated whenever there is an identified market change and according to the development progress of the company's new product. Considering innovation success and the outcome from verification and validation activities, the value actually incorporated into the product might vary from the initial plan.

Table 10.1 Current condition field contents

What is the market?	How is it segmented?
	Do we expect to have product/process variations to different segments?
	How does the customer(s) use the product? (What do they need, what is "value" to each of them)
	Necessary requirements/functions/features of the product? (must haves)
	Additional requirements/functions/features would enhance customer experience? (nice to haves)
Who are the competitors?	Name, picture, price, value chain
	Quality differences
	Other differences
Who are the substitutes	How they fulfill the same needs?
	Name, picture, price, value chain
Who are the main suppliers?	Number, size and location
	Uniqueness of service
	Goals/targets

We recommend using either a graph, chart, or table to make the comparison visual. Radar charts, even though bringing graphical and easy to understand information, lose resolution if the number of value items to compare or products to consider increases, thus becoming difficult to read. When comparing lots of data, tables are a better choice.

In the case of a process development project, this field must include details comparing the process AS-IS and the designed process TO-BE against the identified value items.

10.2.4   Planning for What Is Relevant

The team (even if it is a one-man team) should create an initial version of the milestone chart. This will help them to keep the focus, give priority, and reduce the waste in this phase of the PDP.

Milestone charts are similar to bar charts but only identify the scheduled start or completion of major deliverables and key external interfaces (Fig. 10.3) [3]. This approach helps the team to keep focus and prioritize deliverables. We would rather use this kind of chart since it gives a broader view of what dates cannot be missed and reduces the wastes of wishful thinking, unnecessary processes, scheduled wait, and all the consequences from detailing the whole set of activities for the whole development team.

Note that the milestone chart gives a program-level view of the complete development project, thus facilitating the team meetings. Each individual team shall have a more detailed planning, even using bar charts, to help the planning, execution, and control of their specific work. But at the team level they have much more knowledge of how to detail their exact activities and with minimum wishful thinking.

The minimum set of milestones should include the dates the team expects to have the PD Visual Management Board and the VFD fields filled. The milestone chart is reviewed and updated at each team meeting. If you want more control, you can adopt the EVA (Earned Value Analysis—see Chap. 4) to keep track of the development progression. We strongly suggest, though, using instead a *kanban* system as presented in sequence.

10.2.5   Kanban and Product Development

The progress board use was borrowed from Agile Methods, and it implements the Kanban system into our product development project. After creating the milestone chart, the team has all the key dates set, but no activities defined to reach any of these dates and with the expected results.

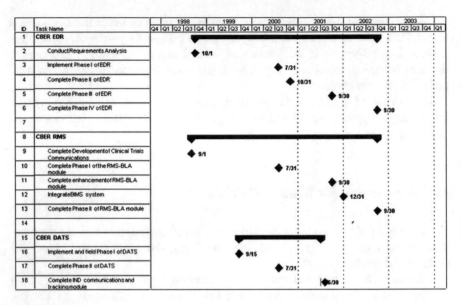

Fig. 10.3   Milestone chart

Before you start the real work, you should decide the periodicity of the team's meetings (we recommend daily or weekly), define the set of activities to be performed in order to achieve the next milestone, and define who is responsible to do what until the next meeting.

Figure 10.4 shows that the set of activities to be performed between milestones shall be put in the "not checked out" space. Whenever a team member starts an activity it is moved to the "checked out" area, and, after completion, to the "done" space. The burn down chart keeps track of the team's progress, by graphically decreasing the amount of work not initiated, which ideally will reach 0 (zero) until the next milestone, when this process will start again.

In each meeting, the Progress Board is updated considering the work accomplished and the tasks still to be performed. It is important to keep track of the team productivity in order to have a good sense of how likely they will finish all the work required until the next milestone. This track of productivity is a good measure of the team's capacity to deliver the development results on time and according to the budget. This field is reviewed and updated in each team meeting.

10.2.6 The Road Ahead Is Always Bumpy

This field is a repository of all identified risks and issues, and the corresponding planned mitigation or corrective measures. Whenever a risk or issue is identified, mitigated, or solved, this field is revisited and updated.

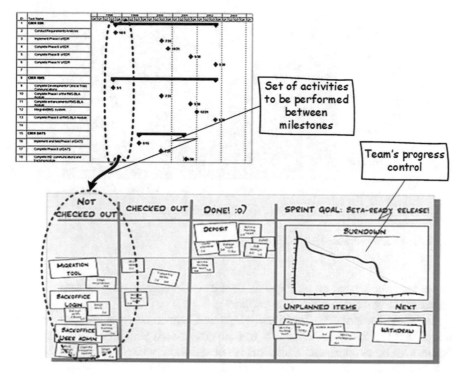

Fig. 10.4 Progress board example

Identifying risks is a tricky job. People often mistake consequences as risks, while the risks are the very causes of these consequences. You can make a parallel to when you go find the root causes by asking why. For instance, having the project go into overtime and/or over budget is a consequence; but why do you believe this might happen? Maybe you believe there will be changes in the exchange rates and you have important product parts which are imported. Maybe you believe you might suffer delays from a supplier. Maybe you are not sure how easily you are going to master a particular technology. One cannot plan mitigation actions to "go into overtime and/or over budget," but you can think of actions that face the particular situations that might cause them. These particular situations are the risks.

After identified, qualitative and quantitative impact analysis must be carried out for each risk. A likelihood vs consequence chart (Fig. 10.5) supports this analysis, by either assigning qualitative or quantitative weights to each of the following attributes and respective weights:

10.2.6.1 Likelihood

Remote (weight 1)—These are things that have a near 0 % chance of happening to you: "You do not live anywhere close to the Pacific Ring of Fire and your area

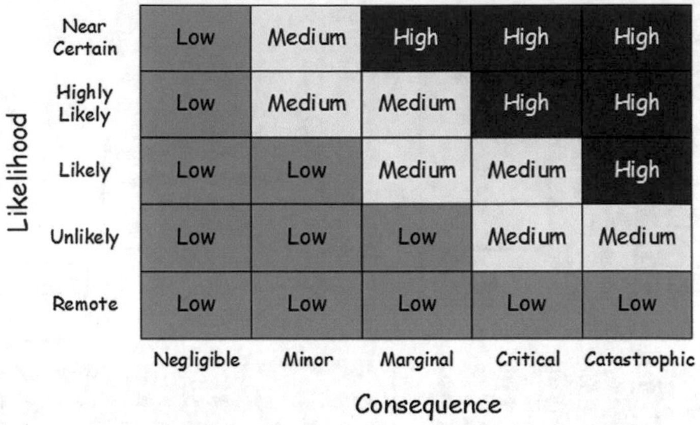

	Negligible	Minor	Marginal	Critical	Catastrophic
Near Certain	Low	Medium	High	High	High
Highly Likely	Low	Medium	Medium	High	High
Likely	Low	Low	Medium	Medium	High
Unlikely	Low	Low	Low	Medium	Medium
Remote	Low	Low	Low	Low	Low

Likelihood (vertical axis)

Consequence

Fig. 10.5 Likelihood versus impact chart

has never experienced an earthquake, then it would come as a complete and total anomaly for you to experience an earthquake."

Unlikely (weight 2)—These are things that happen on a regular basis but are less than 33 % likely to happen to you directly: "Each year 1 in every 34 in 100 homes will be burglarized. While this is not statistically likely to happen to you it is something that happens on a regular basis."

Likely (weight 3)—These are things that are statistically between 33 and 66 % likely to happen to you: "About once every three years you face some electric shortage; it is statistically likely that in any given year you have a 33 % chance of facing electric shortage."

Highly Likely (weight 4)—These are things that are probably going to happen. The statistics are 67 % or better that you will experience it at some point in your life: "Statistics say you will change a burnt light bulb once every year of your life. At least one burnt light bulb in five years is highly likely."

Near Certain (weight 5)—These are things that have happened to you before and are almost certainly going to happen again: "You live in Tornado Alley, you have had tornadoes in your area and will almost certainly see them again."

10.2.6.2 Consequence/Impact

Negligible (weight 1)—This would not cause a hiccup in your routine, budget, emergency fund, food/water storage or comfort.

Minor (weight 2)—This may cause a hiccup in your comfort or routine but would not affect your budget, emergency fund or food/water storage.

Marginal (weight 3)—This would probably interrupt your routine and your comfort, you may have to dip into your food/water storage on a short term or medium term basis and your budget might be disrupted slightly but your emergency fund would not be touched and your budget would not be altered drastically or on a long term basis.

Critical (weight 4)—Your comfort and routine will be drastically altered. Your budget will be altered significantly and on a long term basis. Your emergency fund will be necessary and you will be using your food/water storage on at least a medium term basis.

Catastrophic (weight 5)—Your comfort and routine will be completely destroyed. Your budget will be altered drastically and on a long term basis. Your emergency fund will not be enough to repair the damage and you will be relying on your food/water storage on a long term basis.

The total risk impact equals to its likelihood weight versus its consequence weight. For instance, a risk which is unlikely to happen (weight 2), but has critical consequences (weight 4) results in a total impact 8 (2*4).

10.2.7   Fill the VFD's Value Identification Matrix

The filling of the VFD occurs according to the steps presented in Chap. 9.

10.2.7.1   Stakeholders' Identification

The first step while filling the VFD is identifying the stakeholders. Stakeholders have to be considered regardless of whether they are inside or outside of the development company (Fig. 10.6), or if they contribute directly or indirectly to the development (which is the case of regulatory agencies).

External stakeholders are the ones who pull value from the product development program's final results (the product and/or services). They can be encountered when we consider the "Product/Process Follow-up" and the "Product Discontinuation" process groups from the PDP.

Internal stakeholders, by the other hand, relate to the value chain, and are the ones who pull value from the product development program's intermediate results. They can be encountered when we consider the "Design & Development" and the "Production/Ramp-up" process groups of the PDP.

The following questions support the stakeholders' identification (adapted from [4]). Being a generic questionnaire, it can be adapted to any environment. In order to be considered a stakeholder, at least one of the questions must have a positive answer. In each question, answer "Who _____?"

1. approves the development budget?
2. approves the functional requirements?
3. approves the technical requirements?
4. approves the engineering design decisions?

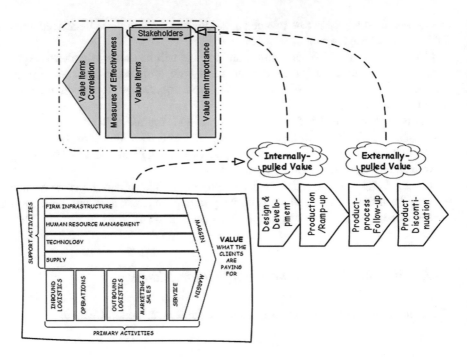

Fig. 10.6 Where are the stakeholders

5. approves requirements changes?
6. approve budget changes?
7. is going to use or interact with the produced product or service?
8. defines the organizational goals that led to the development?
9. is going to allocate people to the development team, and determine the amount of hours per day that they will work?
10. is going to approve the contracts with suppliers?
11. is the development sponsor (who can use his authority supporting the team to overcome organizational obstacles)?
12. is going to manage the development (ensuring that tasks are completed on time and within the budget and that problems are identified and resolved)?
13. represents the organizational policies governing this development?
14. represents the regulations and laws that affect this development?
15. will have their work disrupted by the development?
16. will have to change their work systems or processes due to this development and its results?
17. will benefit from this development?
18. will perform the work (including vendors, subcontractors, besides the company's own employees)?
19. makes the approval decisions for phase change during the development process?

Table 10.2 External and internal stakeholders

External	Internal
User	Shareholder/sponsor
Customer	Suppliers
Shareholder/sponsor	Design and development team
Dealer	Production
Distribution logistics	Development partners
Training network	Quality
Service network	Tests
Recycler	Distribution logistics
Regulatory agencies	Recycler (scrap)
	Regulatory agencies

The stakeholders list, thus, can be quite extensive, including from the end user and external customers who decide to purchase the product, to the internal customers of the product such as the sales force, the customer service department production, and so forth. In order to handle this list, we defined the stakeholder relevance according to the following prioritization criteria:

- **Primary**: define having the right or wrong product (i.e., the customer/client and the final user), or have the ability to cancel the development (i.e., the sponsor and the shareholder);
- **Secondary**: contribute decisively to the development; if not satisfied may cause strong disruptions on the PDP flow, or that might erode the product's image in the market (if developing a process, they can affect the process credibility) and
- **Tertiary**: participate in the development, but have minimum power to impact its flow or results.

Table 10.2 shows some examples of internal and external stakeholders.

10.2.7.2 Analyze the Value Items

During the value items analysis, the PD team shall:

1. Identify the pulled value;
2. Solve the ambiguity from the pulled value items.

10.2.7.3 Identify the Pulled Value

For each identified stakeholder group, the development team must understand the value the stakeholders expect) on their own terms. Several approaches can be used to elicit the value: "go and see," interviews, and so forth. We strongly recommend using the value targeting process and product use analysis (Fig. 10.7), as presented in Chap. 7, so it makes you "stand in the stakeholder's shoes" and understand the challenges they face regarding the product, for instance:

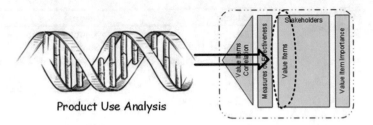

Product Use Analysis

Fig. 10.7 *Genchi genbutsu* DNA

Fig. 10.8 Value flow
diagram

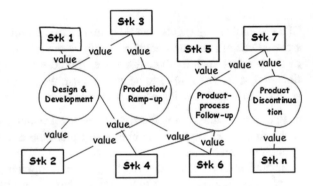

- **Manufacturing**: smooth flow through the line, flexible lines ready, mass customization ready;
- **Distribution**: marketing channel strategy;
- **Selling**: marketing publicity strategy and point of sale;
- **Service/Use and Maintenance**: fulfilling the customer/user's needs, maintenance;
- **End of life/Disposal**: recycle/reuse

To support the value identification and organization, we suggest using a value flow diagram (Fig. 10.8) where the value is pulled by the stakeholders through the PDP's execution and use phases. In order to build the diagram, for each process group you should ask who are the related stakeholders, and for each stakeholder you should ask what does he/she expect from using the product (in whatever state it is: design, prototype, final product, final process, etc.) during each process.

Another tip is identifying the main functions the stakeholder will perform while interacting with the development results, either final or intermediate, in order to perceive its individual benefits and overall value.

While identifying value, special attention must be given to understand all possible dimensions from which the stakeholder might perceive value. We should not be limited, thus, to product/service use (functional aspects), but also consider all the possible needs beyond the product functionalities, some categories of nonfunctional pulled value can be:

Safety—Safety related value for system functions are determined by identifying and classifying associated functional failure conditions. All functions have associated failure modes and associated effects, even if the classification is "No safety effect." Safety related functional failure modes may have either contributory or direct effects upon the product safety.

Security—The security related values cover the security-related areas with regard to protecting the confidentiality, integrity, and availability of the product/service itself and the information processed, stored, and transmitted by the product/service. Depending on the development product (from an individual product or service to a complete value chain), these areas might include: access control; awareness and training; audit and accountability; certification, accreditation, and security assessments; configuration management; contingency planning; incident response; maintenance; media protection; physical and environmental protection; personnel security; communications protection; and information integrity.

Performance—Performance refers to parameters such as range, accuracy, capacity, size, weight, consumption, etc. These are the critical performance parameters necessary for delivering the expected value through the complete product lifecycle.

Use—Refers to hours of operation per day, duty cycle, shutdown routines, a percentage of capacity used, and so forth. To what extent will the product be used at each of its lifecycle's phases? This leads to the determination of the level of stress imposed on the product by anyone who deals with it.

Maintainability—Includes scheduled and unscheduled maintenance needs and any links to specific safety-related functions. Factors such as the percent of failure detection or the percent of fault isolation may also be important. Provisions for external test equipment signals and connections should be defined at this moment.

Distribution—The logistics related values assure the product will be available at the expected distribution places and in the correct amount (or capacity in the case of services).

Reliability—Is quantitatively defined, including cost/effectiveness of the product, operational availability, dependency (coupling with other products), mean time between failures (MTBF) failure rate, readiness rate, maintenance downtime, mean time between maintenance (MTBM), facility use (percent), need for staff and their qualifications, cost, etc.

Certification—This includes additional functions, functional attributes, or implementations that may be required by worthiness regulations or may be necessary to show compliance with worthiness regulations.

Life cycle—Anticipates the product lifecycle. How long will the product be used by the client? What need is there to stock the product and its parts (if any)? Where is the inventory stored?

Environment—Where the product should operate efficiently. Examples include: temperature, shock and vibration, noise, moisture, Arctic or the Amazon, mountainous terrain or plane, airborne, on the ground, boat, space, etc. In which conditions will the product be subjected to throughout its lifecycle and for how

Table 10.3 From value to value items

Value	Value item
1 Realign the aircraft	1.1 Quick response to triggering
	1.2 Return to normal flight attitude
	1.3 Eliminate aerodynamical effects on the aircraft after use
	1.4 Eliminate electrical effects on the aircraft after use

long? In addition to issues related to the operation, the environmental pulled value should consider ways of shipping, handling, and storage (it is possible that the product is subject to stricter conditions during transport than during operation).

10.2.7.4 Solve the Ambiguity From the Pulled Value Items

Finally, since the value initially understood from the stakeholders might not be clear, the development team must work on clarifying that into unambiguous value items. Ambiguity means the existence of multiple conflicting interpretations of the information held and required which leads to a lack of consistent information. Solving ambiguity leads to the search for the meaning of things.

The work of ambiguity elimination is the exact work of requirements engineering. Therefore, the initial pulled value must be drilled into value items, which are very similar to user requirements, once they are written from the stakeholder's point of view. User requirements define the information or material that is input into the business process, and the expected information or material that is the outcome from interacting with the business process (system), specific to accomplish the user's business goal.

The value as pulled by the stakeholders is further detailed into value items, by asking "What do you mean by that?" Table 10.3 shows the final set of value items defined from the initial pulled value of [1. Realign the aircraft]; note that the actors were suppressed in the VFD, since the value delivery matrix links stakeholders (actors) to value items.

10.2.7.5 Prioritize the Value Items

Each considered stakeholder has particular needs, thus rating differently the value items importance. Also, as already stated, each particular stakeholder has different relevance. The value items prioritization takes into account the combination of these ratings. Therefore, the importance of the value item VIi is obtained as:

$$\text{IMP}_{\text{VIi}} = \sum_{j=1}^{k} \text{SR}_j * \text{IS}_j \tag{6}$$

where:

- **SRj =** is the relevance of the jth stakeholder, ranges from 9, 3, and 1, if the stakeholder was considered as primary, secondary or tertiary, respectively:
- **ISj =** is the interest from the jth stakeholder on the value item VIi, ranging from 9, 3, 1, and 0, on the case of high, medium, low, and not important for that particular stakeholder, respectively:

When rating the interest to stakeholder consider:

- High: the item falls into a *must have* category.
- Medium: the item falls into a *nice to have* category.
- Low: the item relates to the stakeholder (he can perceive it), though he does not care about it at first.
- None: the item does not have any relation to the particular stakeholder.

Table 10.4 shows the stakeholders and a value items subset from the stall recovery system example, where the linking between the [Client] and the value item [1.1 Quick response to triggering] contributing with "81" (primary * high = 9 * 9). By repeating this calculus trough the line, the total importance of this item is 360 (three hundred and sixty).

10.2.7.6   Define Measures of Effectiveness (Moe)

A PDP provides no value if it does not have the capabilities required by the end customer. These capabilities must be translated into identifiable and measurable parameters that can be designed, developed, and tested.

In order to verify the presence of the value items in the development results (product and/or services), at least one measure of effectiveness (MoE) must be defined for each value item.

For instance, the measure of effectiveness for the value item [1.2 Return to the normal flight attitude] was [return the aircrafts' angle of attack (AoA) to $M \pm D$ degrees in less than T seconds].

Since we are talking about MoEs for value items, which are very similar to user's requirements, these measures must be defined in a way that the related stakeholders could measure it themselves (perceive while using the product/process).

10.2.7.7   Identify Conflicting Value Items

Conflicting value items are items that cannot be optimally delivered simultaneously (like having a robust and fail proof product, while aiming to a minimum mass) if using the current company knowledge and capacity.

The conflicting value items direct the creation of trade-off curves that, besides aiding the development team, are part of the company's knowledge assets. By

Table 10.4 Value item importance

Value	Value Item	Client	Enterprise - Owners	Enterprise - Business Unit	Enterprise - Quality	Enterprise - Industrial Engineering	Enterprise - Production	Enterprise - Homologotion	Enterprise - Commercial Department	Enterprise - Logistics	Suppliers - Piece and Parts	Suppliers - Labs and Testing	Importance
		Pri	Pri	Pri	Sec	Sec	Pri	Sec	Sec	Sec	Ter	Ter	
1 Realign the aircraft	1.1 Quick response to triggering	h	m	h	m	m	m	m	m	m	l	l	360
	1.2 Return to normal flight attitude	h	m	h	m	m	m	m	m	l	l		360
	1.3 Eliminate aerodynamical effects on the aircraft after use	m	l	l	l	l	l	l	l	l	l		108
	1.4 Eliminate electrical effects on the aircraft after use	l	l	l	l	l	l	l	l	l	l	l	90

Table 10.5 Stall recovering system conflicting value item

Primary value		Secondary value
Return (quickly) to normal flight attitude	YET	Have low mass

challenging and improving the trade-off curves, a company becomes more competitive. These curves, besides supporting decision making, are a simple and convenient way to make reusable knowledge.

Table 10.5 shows one example of conflicting value items, in the context of the stall recovery system. The "primary value" is the value item that has higher importance, in the case of a trade-off they are the ones to be kept a higher values.

 10.3   A Practical View

Although this chapter has been mainly practical, some additional tips can also be given:

- Take some time to prepare yourself for the value identification: read the background, the current condition, and the risks already identified. Value identification is always an opportunity to challenge previously developed information. Understand all types of pulled values.
- Whenever possible apply the go and see to identify the value, through go and see it is possible to "feel" the real voice of the client. It is an opportunity to perceive important value that the stakeholder would not verbalize and to understand the real priority. When asked, stakeholders tend to request what they *like* rather than what they *need*.
- We know we can't avoid interviews, but try to use them as a confirmation technique, and always avoid direct questions and questions which answers are monosyllables (like YES and NO).
- Think simply about the measures of effectiveness. At this moment, we are at the business/user requirements level, therefore the MoE should be stated on terms they understand and are able to check in their environment, without fancy testing tools and procedures.

It's also important to highlight the difference we consider between value and requirements. Value is described much more in the stakeholder's language. Even after having the ambiguity reduced and detailed into value items, these items must be easily understandable by the stakeholders who pulled them, while requirements might be written in a way to facilitate the understanding for the product development team.

When we combine the value items with its measures of effectiveness, we have almost the equivalent from system requirements, once we rather describe how the final product/service should interface with its users, than how the product/service architecture is going to be. By having the value item + MoE, a requirement-like criteria is attained:

Necessary. The stated value item is necessary, once it is pulled by one or more stakeholder.

Concise (minimal, understandable). The value item statement includes only one value item stating what must be delivered and only what must be delivered, stated simply and clearly. It is easy to read and understand.

Implementation free. The value item states what is required, not how the value item should be met.

Attainable. (achievable or feasible). Be defining the MoE, it is possible to imagine which validation and verification strategies can be used to confirm that the value is present into developed product/service.

Unambiguous. Each value item must have one and only one interpretation. Language used in the statement must not leave a doubt in the reader's mind as to the intended descriptive or numeric value.

Some other requirement criteria are not applied to value items. During the execution phase the value items are going to be detailed into requirements, never losing the traceability to the value items they came from. When detailed into requirements the following criteria should also be attained:

Complete. (standalone) The stated requirement is complete and does not need further amplification. Requirements should be stated simply, using complete sentences. Each requirement paragraph should state everything that needs to be stated on the topic and the requirement should be capable of standing alone when separated from other requirements.

Consistent. The stated requirement does not contradict other requirements.

Once the pulled value has been identified and prioritized, the next step is defining how the value set is going to be delivered and understanding the risks for doing so, which is the subject of the next chapter.

References

1. Blanchard BS, Fabrycky WJ (1988) Systems engineering and analysis, 3rd edn. Prentice Hall, New Jersey
2. Haik Y, Shahin TM (2003) Engineering design process hardcover, 3rd edn. Cengage Learning, Stanford
3. Project Management Institute, PMI (2013) A guide to project management body of knowledge (PMBOK® Guide), 5th edn. Project Management Institute, Newton Square
4. Verzuh E (2005) Stakeholder management strategies applying risk management to people. In: Proceedings of PMI Global Congress North America, Toronto. Project Management Institute, Newton Square

Chapter 11
Study Phase—Value Proposition Activities

While the value identification activities were concerned about understanding and structuring all the value pulled by the stakeholders, the next step in the PDP (Fig. 11.1) aims to develop a possible functional architecture that can deliver this value, which is the chief engineer's vision of the upcoming product/service. The value proposition documents in this vision differ from the initial pulled (product) vision. This value proposition is deeply rooted in the identified value set, and defines which product/service functional architecture is the preferable choice to deliver this set and yet mitigate its related risks, therefore guaranteeing the PDP uninterrupted flow. This chapter uses the stall recovery system project example to present a stepwise execution of this phase's activities, where special emphasis is given to defining the best candidate product's functions to SBCE.

11.1 Introduction

After defining the scope (the value to be delivered), traditional PD chooses a possible product alternative to implement it. They work on it until something fails, requiring loopback rework, which often causes changing to another solution. To avoid these wasteful loopbacks, the lean way, as stated in Chap. 9, avoids choosing just one alternative and uses SBCE to maintain and streamline multiple alternatives during the product development. Through SBCE variations or modifications, an existing product or service (incremental innovation) and completely new ideas (breakthrough innovation) can coexist.

Even though the SBCE brings several benefits, carrying out several alternatives to each product's subsystems requires having the necessary people and investment to allow concurrent work on those alternatives.

In order to define a good (and not too costly) strategy to explore the solution space, the LPDO must cluster the value items into functions and access the risks to deliver each of these functions. We suggest that by knowing the amount of value and the related risks, you can identify the most critical functions which are the

© Springer International Publishing AG 2017
M.V.P. Pessôa and L.G. Trabasso, *The Lean Product Design and Development Journey*, DOI 10.1007/978-3-319-46792-4_11

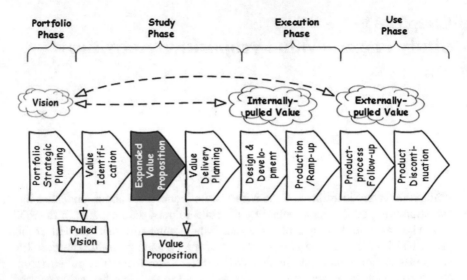

Fig. 11.1   Value proposition activities in the PDP

best candidates to SBCE. Indeed, a functional subsystem, with higher value delivery potential and which the risks to successfully delivering this value are also perceived as high, would benefit more from applying the SBCE.

11.2   Using the Board to Guide the Value Proposition Activities

During the value identification activities we filled in most of the PDVMB, which will be used from now on as a combination of working board and management cockpit.

As you advance further in the PDP activities, constant crosschecking among the PDVMB and VFD is paramount, since they provide an overall and concise picture which is the greatest benefit from going visual:

- **Positioning sense**: The background, current condition, comparison of products, and the VFD matrices fields give you a clear view of the vision where you want to be, where you actually are, and how you compare to your current product, competitors, and substitutes. Any decision that deviates from these "positioning instruments" is wasteful. Changes from the information here must be previously negotiated with the related stakeholders, and might be supported from additional (go and see) facts.
- **Progress sense**: The milestone chart and progress board provide a sense of achieving and progressing towards the development goals achievement.
- **Uncertainty sense**: The risk and issues field gives you a good sense of how clear the path is ahead, and also gives you the means to act proactively.

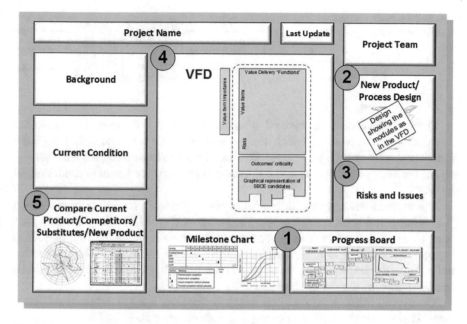

Fig. 11.2 PD visual management board filling sequence

The PDVMB and VFD filling sequence described below (Fig. 11.2) will guide you during the value proposition activities. Note that at this moment the focus changes to the VFD's Rework Avoidance Sub-matrix.

11.2.1 Milestone Chart and Progress Board

Both the milestone chart and the progress board must be updated and reviewed at each team meeting. Activities will progress from Not Checked Out ≫Checked Out ≫Done through the progress board, and the team shall control the work pace according to this progress, also considering any risk mitigation and issues solving activities added to the backlog.

11.2.2 New Product/Process Design: Functional Architecture

Once the value items set has been identified, which defines the problem the PDP must solve, we begin the solution/product design effort. A good way to start from this point is to identify the main function at the system level, by checking the value pulled by the stakeholders while interacting with the product/service itself (see the value flow diagram in the previous chapter).

Fig. 11.3 A black box that
widens the system boundaries

The functional analysis method offers such a means of considering essential functions and the level at which the problem is to be addressed. The essential functions are those that the product, system, or service to be designed must satisfy, no matter what physical components, service processes, or business model might be used [1, 2].

The method starts by considering the whole product, system, or service as a "black box," and identifying its inputs and outputs (Fig. 11.3). The black box contains all the functions that are necessary to convert the inputs into outputs. It is paramount to ensure that all the relevant inputs and outputs consider the value pulled from the stakeholders. As a rule of thumb, the black box function should be broad—widening the system boundary.

The overall (black box) function is now broken into a set of essential sub functions. There is no single way of doing that, but one must consider the process steps that transform the inputs into outputs. Do not forget that there might be more than one process stream, so that all the input and output relations are solved. In the same way, the value pulled by the stakeholders will give good support to executing the functional breakdown.

Since we are considering a broad value set, some of the identified essential functions might not be directly related to the product itself, but to services that are necessary to deliver the value, or even a complete new value stream.

Draw a "white box" diagram (Fig. 11.4) to make visual the relations of the sub functions among themselves, and the inputs/outputs. This diagram should be placed in the board's "New Product/Process Design" field in order to give reference to further discussions about the product, and will remain there until the product architecture is defined during the execution phase. The diagram shall be updated whenever a change in the functional architecture occurs.

In the case of a process development project, this field must present the functional sequence for the process TO-BE. In the same way that it has happened in the case of the development of a product, the process TO-BE might differ considerably from the process AS-IS, which might even have been split into more than one process.

If during the product development process, you consider delivering both products and services (a product and related service packages, for instance), you might have several input-output functions sequences to describe each product and/or service.

The functions to be considered are either performed by the product or part of the offered service and not an action or process executed by any of the

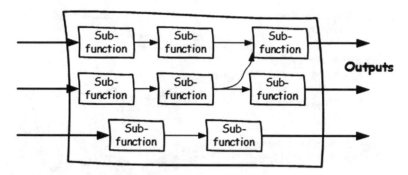

Fig. 11.4 A white box diagram with all the essential sub functions

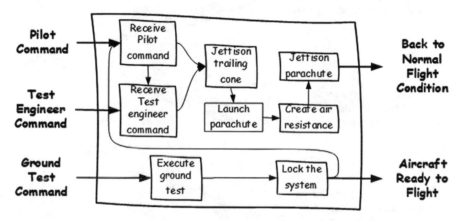

Fig. 11.5 Identified functions from the stall recovery system example

stakeholders. A common mistake is making a sort of process mapping, having the client or any external stakeholders as an actor.

Considering the stall recovery system example (Fig. 11.5), we identified functions to (1) recover the aircraft from abnormal flight condition, and (2) make sure that the system is safe and ready for flight.

This functional architecture is proposed and developed in such a way that the stall recovering system will work according to what is presented in Fig. 11.6. The trailing cone is used during flight tests; it's a device at the end of a cable which has a pressure sensor partway along it where it can make static pressure measurements in the free air away from disturbances created by the aircraft.

Defining the functional architecture is sometimes tricky; we suggest also checking the examples from Chaps. 15 and 16 before doing your first try.

Normal flight; system on
standby

Risky maneuver; system
armed

Out of control aircraft

Deploy command activated;
1 - Trailing cone jettisoned

Deploy command activated;
2 - Parachute launched

Parachute opens to
stabilize aircraft
(nose down)

Jettison command
activated; parachute
cable is cut

Aircraft back to
normal flight

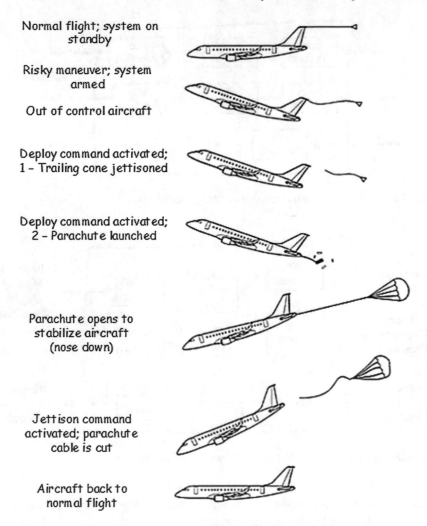

Fig. 11.6 The stall recovering system working

11.2.3 Risks and Issues

During the first risk assessment round made as part of the value identification activities, the work was mainly based on background information. Now that you have a better knowledge of the product/service to be produced you can dig deeper into the actual risks.

We suggest a complete and comprehensive risk assessment both for the development program as a whole and for each function related to the product. Note that for each line on Table 11.1 the development team shall identify whether there is an uncertain event or aspect related to the development program, which, by happening or being present, might impact the development project. Qualitative and quantitative impact analysis must also be carried out for each identified risk, as presented in the previous chapter.

Topics 1, 2, 3, and 4 from Table 11.1 relate to the development process as a whole, while Topic 5 relate to each identified function separately. We do not expect that you identify risks for every line of the table, but that you consider possible (if any) risks from each line. Therefore, it is not a problem if you identify risks from only some of the presented aspects.

11.2.4 Fill the VFD's Value Rework Avoidance Sub-matrix

This phase's main objective is filling the VFD's Rework Avoidance sub-matrix, which summarizes in a very concise and visual way the value proposition, by linking value items and program risks to the product's value delivery functions. Through this linking, the function's criticality is calculated and the SBCE need is estimated. Therefore, this sub-matrix guarantees not only the full coverage among functions, value items and risks, but also suggests which functions would greatly benefit from SBCE.

The following VFD filling occurs according to the steps presented in Chap. 9.

11.2.4.1 Define the Value Delivery Functions

All the sub-functions (functional modules) identified during the functional analysis shall be transported to the VFD, and related to the value items they support delivering (Fig. 11.7). Each function must support the delivering of at least one value item, and all the value items from the complete value items set must be addressed by at least one function. If you find something different from that, you must check the essential sub functions for unnecessary or missing functions.

The function of the effective value item delivery contribution is then rated as high, medium, low, or none, corresponding to the weights 9, 3, 1 and 0, respectively:

- **High (9)**: The function plays a critical/central role to deliver the value item, and without it the value item is absent from the product.
- **Medium (3)**: The function plays a supporting role in delivering the particular value item by the high rated functions.
- **Low (1)**: The function interfaces to a high rated value delivery function, and might indirectly influence the value item delivery.
- **None (0)**: The function has no relation either to the value item itself or to any function which is highly rated in delivering this value item.

Table 11.1 Program risk assessment aspects

1. Economic	1.1 Inflation	1.1.1	Workforce
		1.1.2	Materials
		1.1.3	Equipment
		1.1.4	Services
	1.2 Financial Uncertainty	1.2.1	Contractor
		1.2.2	Development company itself
		1.2.3	Suppliers
		1.2.4	Financial institutions
		1.2.5	Exchange rate fluctuations
	1.3 Market	1.3.1	Competitors
		1.3.2	Demand change
2. Contract	2.1 Lack of payment	2.1.1	To Suppliers
		2.1.2	From Contractor
	2.2 Delays	2.2.1	Own company
		2.2.2	Contractor
		2.2.3	Suppliers
	2.3 Contract changes	2.3.1	Contractor
		2.3.2	Own company
	2.4 Workforce disputes	2.4.1	Disputes (union)
3. Politics	3.1 Environment	3.1.1	Damages to environment
		3.1.2	Social risks
		3.1.3	Natural disasters
	3.2 Government acts	3.2.1	Change on interest rates
		3.2.2	Law changes
		3.2.3	Regulation changes
		3.2.4	Tax legislation
		3.2.5	Labor law
		3.2.6	Patents and licenses
		3.2.7	Embargoes
4. Execution	4.1 Workforce uncertainty	4.1.1	Availability
		4.1.2	Ability
	4.2 Equipment uncertainty	4.2.1	Availability
		4.2.2	Quality
		4.2.3	Operability
		4.2.4	Integration
	4.3 Materials uncertainty	4.3.1	Availability
		4.3.2	Preservation
		4.3.3	Storage
	4.4 Productivity	4.4.1	Delays
		4.4.2	Variation (quantity)
		4.4.3	Quality
		4.4.4	Safety
		4.4.5	Managerial competence
		4.4.6	Manufacturability

Table 11.1 (continued)

5. Technical (for each identified function)	5.1 Product	5.1.1	Complexity
		5.1.2	Maturity
		5.1.3	Dependency
		5.1.4	Performance
		5.1.5	Technologies availability
		5.1.6	Innovation
	5.2 Logistics	5.2.1	Testability
		5.2.2	Maintainability
		5.2.3	Reliability
		5.2.4	Customer service

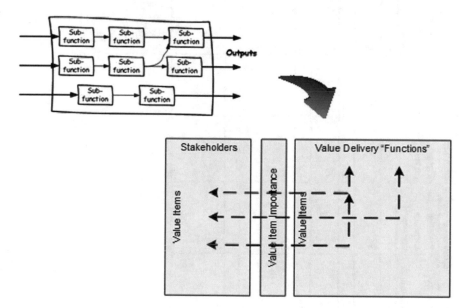

Fig. 11.7 Relating functions and value items

Table 11.2 exemplifies this correspondence and the distribution of the total value item importance among the related functions, which means how much each function contributes deliver it. The value item [1.1 Quick response to triggering], for instance, is delivered by the cooperation between seven different functions, two of them with high contribution ($2*9 = 18$), three with medium contribution ($3*3 = 9$), and two with low contribution ($2*1 = 2$). The total item's importance (360) was divided into 29 shares ($18 + 9 + 2$). Therefore, each high, medium, and low contribution function received 112 ($360/29*9$), 37 ($360/29*3$), and 12 ($360/29*1$) out of 360, respectively.

Table 11.2 Distributing the value item importance among the contributing functions

Value	Value Item	Importance	Receive pilot command	Receive test engineer command	Create air resistance	Launch parachute	Jettison trailing cone	Jettison parachute	Lock the system	Execute ground test
1 Realign the aircraft	1.1 Quick response to triggering	360	h 112	m 37		h 112	m 37	m 37	l 12	l 12
	1.2 Return to normal flight attitude	360	l 18	l 18	h 162	m 54	m 54	m 54		
	1.3 Eliminate aerodynamical effects on the aircraft after use	108	h 36	l 4	l 4		m 12	h 36	l 4	m 12
	1.4 Eliminate electrical effects on the aircraft after use	90	h 39	h 39						m 13

11.2.4.2 Address Risk Response

Identify the risks to be actively mitigated according to the risk acceptance limits
set for the project. As a rule of thumb, all the risks that receive a high risk rate on
the likelihood impact chart (Fig. 11.8) shall be considered active mitigation, other
risks might also be included, depending on the strategy.

Be aware that low rated risks might have complex relations among themselves,
and the occurrence of a combination of them might have a relevant impact, or even
the occurrence of one of them might increase the likelihood of another. This is the
reason why the continuous monitoring of the risks from the risk set shall be done,
even if you do not include all the risks in the VFD at this moment.

Now the risks that the team decided to mitigate shall also be filled into the
Rework Reduction Sub-matrix and linked to the related functions. Each risk must
be associated with at least one of the value functions, but the contrary is not true,
since you can have functions which have no risks either identified or elected to
active mitigation. Necessary adjustments to the function set will be made until the
correspondence among it and the risks is complete since the risks can help you to
identify a function that might have been forgotten.

Similar to the value items importance, the risk impact is distributed among the
related functions. This relation can be rated as high, medium, low, or none, corresponding to the weights 9, 3, 1 and 0, respectively:

- **High (9)**: The risk has a critical/central negative impact on developing the function, therefore having the risk means not having the function.
- **Medium (3)**: The risk has an important negative impact on developing the function, therefore having the risk means not having the function in its fullness.
- **Low (1)**: The risk has some negative impact on developing the function, therefore having the risk means having some minor losses on the function capacity to
  deliver full value.
- **None (0)**: The risk has no impact on developing the function.

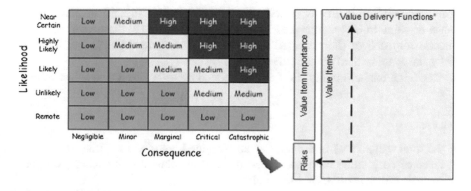

Fig. 11.8 Moving risks to the VFD

Table 11.3 Distributing the risk impact among the potentially affected functions

Identified Risks								
R.1 Parachute does not fit into the mortar	12		h	m				
			9	3				
R.2 Delays on parachute delivery	12		h					
			12					
R.3 Not jointly achieve the release speed and the maximum retreat strength specifications	16		I	h				
			1.6	14,4				
R.4 Do not master the gas generator technology	12			h				
				12				
R.5 Requirement change requests from client	12		m	h			h	
			1,7	5,1			5,1	
R.6 Not be able to buy the mortar MP in time	12			h				
				12				
R.7 Pyrotechnical components validity does not meet the lifetime and/or maintenance requirements	12			m	m	m		h
				2	2	2		6

The total perceived risk value related to a function is calculated by the simple sum of the weights given by the risks that impact it.

Table 11.3 exemplifies this correspondence and the distribution of the perceived risk among the potentially affected functions. The risk [R.1 Parachute does not fit into the mortar], for instance, relates to two different functions, one of them with high impact (1*9 = 9) and the other with medium impact (1*3 = 3). The total risk impact (12) was divided into 12 shares (9 + 3). Therefore, the highly impacted function received 9 (12/12*9), and the medium impacted function 3 (12/12*3) out of 12, respectively.

11.2.4.3 Calculate the Criticality of Each Value Delivery Function

The functions' criticality is directly proportional to: (1) the amount and importance of value to be incorporated in these functions; and (2) the perceived risk to successfully deliver the expected value subset. The more valuable and the more risky, the more critical are the functions.

The functions' criticality (FC) was calculated according to the equation below:

$$FC = TV * PR/100 \tag{11.1}$$

where:

- the total value (TV) is the sum of all weighted contributions (the total importance of each value item was proportionally distributed to the teams, weighted by their effective contribution); and
- the perceived risk (PR) is the simple sum of the weights given by the risks that impact it.

Table 11.4 shows some calculated criticality values for the stall recovery system example. The [receive pilot command function], for instance, had TV = 435, TR = 1, resulting in FC = 4. The most critical function was [launch parachute], scoring FC = 352. Even though it was not the function with higher TV (second to the function [create resistance]), the risks related to it, contributed decisively on the FC.

11.2.4.4 Define the Priority to Parallel Development

Considering a limited budget and the needed return of investment, not all the functions may be developed through SBCE. The functions to be developed through a set of alternatives should be chosen considering the restrictions imposed on the development project and the previously calculated criticality, where the functions that deliver more value and/or are more risky to be successfully developed are the best candidates to SBCE.

Having chosen the SBCE candidate functions, the team should explore the solution space for possible alternatives to implement these functions. One alternative to be promptly considered is how you already perform this function, if any. Other alternatives can be found by applying techniques such as morphological chart and "7 ways."

The alternatives that survive the first drilling against the value items are then transported to the VFD. As a consequence, the functions chosen to SCBE will have more than one column in the VFD, each one considering one of the picked alternatives.

11.2.4.5 Morphological Chart [1, 2]

The morphological chart is a tool used for solving design problems in the field of engineering design. On the morphological chart, the product functions and different mechanisms which can be used to perform these functions are listed. In order to increase the visual impact, the mechanisms can be drawn (pictograms or symbols instead of words) in a way that helps imagining the combined implementation among different combinations of them.

By means of the morphological chart, ideas are generated and combined into several feasible designs using different mechanisms. In other words, each combination of mechanisms suggests a solution to the problem. The generation of solutions is thus a process of systematically combining already existing components.

In point-based concurrent engineering you may be tempted to choose the "safe" combinations of components, but through SBCE you can challenge yourself by making counter-intuitive combinations of components. Once you begin to combine the mechanisms, you'll gain momentum and start to generate lots of ideas, organize your thoughts, change, erase, scrap, and retrieve from the bin.

Table 11.4 Stall recovery system's outcomes criticality

Value	Value Item	Importance	Receive pilot command	Receive test engineer command	Create air resistance	Launch parachute	Jettison trailing cone	Jettison parachute	Lock the system	Execute ground test
1 Realign aircraft	1.1 Quick response to triggering	360	h 112	m 37		h 112	m 37	m 37	l 12	l 12
	1 2 Return to normal flight attitude	360	l 18	l 18	h 162	m 54	m 54	m 54		
	1.3 Eliminate aerodynamical effects on the aircraft after use	108	h 36	l 4	l 4		m 12	h 36	l 4	m 12
	1.4 Eliminate electrical effects on he aircraft after use	90	h 39	h 39						m 13
Identified Risks	…	…	…	…	…	…	…	…	…	…
	R.1 Parachute does not fit into the mortar	12			h 9	m 3				

(continued)

Table 11.4 (continued)

Value	Value Item	Importance	Receive pilot command	Receive test engineer command	Create air resistance	Launch parachute	Jettison trailing cone	Jettison parachute	Lock the system	Execute ground test
	R.2 Delays on parachute delivery	12			h 11			l 1		
	R.3 Not jointly achieve the release speed and the maximum retreat strength specifications	16			l 1.6	h 14,4				
	R.4 Do not master the gas generator technology	12				h 12				
	R.5 Requirement change requests from client	12			m 1.7	h 5,1			h 5.1	
	R.6 Not be able to buy the mortar	12				h				
	MP in time					12				

(continued)

Table 11.4 (continued)

Value	Value Item	Importance	Receive pilot command	Receive test engineer command	Create air resistance	Launch parachute	Jettison trailing cone	Jettison parachute	Lock the system	Execute ground test
	R.7 Pyrotechnical components validity does rot meet the lifetime and/or maintenance requirements	12				m	m	m		h
						2	2	2		6
	Total Value (TV)		435	500	832	726	651	638	429	204
	Perceived Risk (PR)		1	1	23	49	2	3	5	6
	Final rrV*PR/100		4	5	192	352	13	16	22	12

Table 11.5 shows one example of idea generation in the stall recovery system context. You can create possible product architectures by combining different mechanisms for each function. By considering the calculated functions' criticality, you should create more than one mechanism for at least the [launch parachute] function.

Table 11.5 Morphological analysis for the stall recovery system

	Option 1	Option 2	Option 3	Option 4
Receive pilot command	voice	electrical	radio	mechanical
Receive test engineer command	voice	electrical	radio	mechanical
Create air resistance	parachute	vectored push		
Launch parachute	coil	explosive	external pull	Pneumatic
Jettison trailing cone	cut	disengage	acid	Laser
Jettison parachute	cut	disengage	acid	Laser
Lock the system	cut energy	mechanical lock	logical lock	
Execute ground test	built in	external equipment		

11.2.4.6 7 Ways

In the 7 Ways technique, for each critical function chosen to SBCE, 7 different solutions on how to accomplish them are generated. Each alternative is then evaluated against the related value items to identify a possible positive or negative correlation, or even no correlation at all.

The number 7 is a reference goal to challenge the design team to produce a high number of possible solutions to implement a given function. It´s up to the development team members to select the creativity design method or tool that better fits their way of working.

The scores can be summed up using a Pugh analysis (positive [1], negative [−1], or neutral [0] correlation) and the alternatives that best meet the customer needs are candidates to go forward for further refinement. Table 11.6 shows one example in the stall recovery system context, with possible seven ways to be considered for the [launch parachute] function, which was ranked the most critical one:

Table 11.6 Launch parachute function alternatives by using 7 ways

Launch parachute highly related value items and functions	1	2	3	4	5	6	7
1.1 Quick response to triggering	0	1	-1	1	0	−1	0
2.2 System unavailability detectable at the ground	0	1	0	1	0	1	0
2.3 Work when required (reliability)	0	1	0	1	0	1	0
3.1 Mass	0	1	1	0	0	0	−1
3.2 Mechanical interface (do not damage the aircraft)	1	0	1	1	1	0	1
3.5 Environmental conditions	0	1	−1	0	0	1	0
4.1 Post deploy repair < X	−1	1	1	−1	−1	1	−1
4.2 Corrective maintenance time < Y	−1	1	1	−1	−1	1	−1
4.3 Endure Z years in storage	0	−1	1	1	0	−1	0
5.1 Technical viability	−1	1	1	1	−1	0	−1
5.2 Financialand commercial viability	1	0	1	0	1	0	0
5.3 Strategic alignment	0	1	0	0	0	0	0
5.6 Finish according to schedule	0	0	0	0	0	0	0
5.3 High quality of the supplier	0	1	0	1	0	0	0
7.1 Design for manufacturing and assembly	0	0	0	0	0	0	0
7.4 Defined product and process acceptance criteria	0	0	0	0	0	0	0
R.3 Not jointly achieve the release speed and the maximum retreat strength specifications	−1	1	−1	1	−1	0	−1
R.4 Do not master the gas generator technology	0	−1	0	0	0	−1	0
R.5 Reguirement change requests from client	0	0	0	0	0	0	0
R.6 Not be able to buy the mortar MP in time	0	0	0	0	0	0	0
Final Sum	−2	9	4	6	−2	2	−4

1st way: Use the potential energy from a coil system;
2nd way: Use an explosive;
3rd way: Use a device to pull the parachute from the exterior;
4th way: Use a pneumatic system (like an air gun);
5th way: Use the potential energy from tensioning elastic;
6th way: Make a chemical reaction; or
7th way: Use a catapult system.

From the results, the use of explosive and the use of a pneumatic system are the most promising one, and could be carried to SBCE.

 11.3   A Practical View

Some teams focus on quickly determining preferable product architecture, restraining the engineers' imaginations and exploring a limited area from the solution space.

We recommend investing your time the beginning of the development project to explore deeper into the solution space, by letting the engineers' imaginations go free. Forget all the constraints, prerequisites, and limitations and try to build up some perfect/dream alternatives. After doing that, then you start bringing the reality in, with all its limitations. Before pruning your dream due to one of these limitations, try hard to challenge this constraint. Whenever possible, use SBCE to maintain some of your crazy solutions, thus giving yourself the chance to try and innovate. Another tip is checking to see if the identified conflicts among value items were explored while defining the alternatives to SBCE.

In the stall recovery system example, imagine that you have a chance (even slight) of creating a magnetic field that would bring the aircraft back to normal flight. Wouldn't it deserve a try? Particularly if other products from your company could benefit from the gained knowledge.

Also, there is no magical number of alternatives to carry out. You should ponder both the value/risk ratio among the alternatives you are considering and the necessary resources availability to carry out these alternatives. It is even worse to carry out several alternatives with insufficient resources than carrying just a couple of them but with full resource commitment.

We once heard, from a research lab director of an important aerospace company, that they build planes and missiles due to the actual technology limitations, but if they could make lightning come from the skies and destroy the target they would do that; and they never limit their research for this and future possibilities.

References

1. Haik Y, Shahin TM (2003) Engineering design process hardcover, 3rd edn. Stamford, Cengage Learning
2. Cross N (1995) Engineering design methods, 2nd edn. Wiley, New York

Chapter 12
Study Phase—Planning Activities

Planning (also called forethought) is the process of thinking about and organizing the activities required to achieve a desired goal. The planning activities (Fig. 12.1) define the value proposition delivery strategy, which aims to develop the scoped product in the most efficient way, with no/minimum waste, unevenness, or overburden. This chapter uses the stall recovery system project example to present a stepwise execution of this phase's activities, where special emphasis is given to defining the pull events that shape the future development execution.

12.1 Introduction

In order to support a JIT-like execution phase, the development plan is indeed composed by a sequence of "pull events" that guide the development team through the development activities.

Instead of pushing scheduled activities, which themselves push information and materials through the development process, pull events guarantee the value flow, make quality problems visible and create knowledge. They are typically tied to physical evidence of progress, such as: (1) integration events that create "boundary objects" as built engineering projects, mockups, prototypes, etc.; (2) successful endings of checks and validations, which are moments of reducing uncertainty and risk in the program. The pull events set creates a "ladder" where in each step up we get closer to the development success.

According to the Project Management Institute [1], planning consists of those processes which establish the total scope of the project effort, define and refine the objectives, and develop the course of action required to attain those objectives.

Indeed, the PD planning takes place during the whole study phase, where the PDVMB and the VFD create a backbone to support and align the planning activities. By finishing the study phase, we can observe that all the project management process groups [1] have been somehow considered (Table 12.1).

© Springer International Publishing AG 2017
M.V.P. Pessôa and L.G. Trabasso, *The Lean Product Design and Development Journey*, DOI 10.1007/978-3-319-46792-4_12

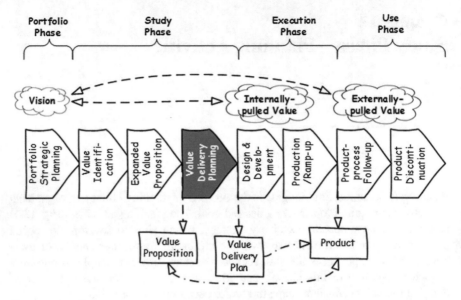

Fig. 12.1   Value delivery planning activities position in the PDP

Table 12.1   Project management process and the book's method

Project management process group	Support to process execution
Integration	Both the VFD and the visual management board
Scope	Visual management board's background field VFD's value identification matrix
Time	Visual management board's milestone chart and progress board VFD's flow definition sub-matrix
Cost	Visual management board's elements balancing VFD's Value identification matrix
Quality	VFD's value identification matrix and flow definition sub-matrix
Human resource	VFD's concurrent engineering sub-matrix
Communications	Visual management board itself
Risk	Visual management board's risks and issues field VFD's rework avoidance sub-matrix
Procurement	VFD's rework avoidance sub-matrix
Stakeholder	VFD's value identification matrix and flow definition sub-matrix

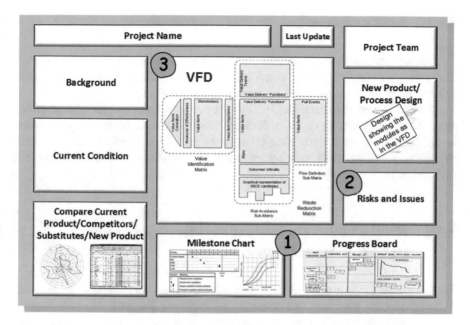

Fig. 12.2 PD visual management board filling sequence

12.2 Using the Board to Guide the Value Delivery Planning Activities

Once having defined the value proposition, your next step is creating a strategy on how to deliver it in the most efficient way. On the PDVMB and VFD filling sequence described below (Fig. 12.2), you are going to focus on defining both the complete PD team and the pull events that will be used during the execution phase of the PDP. In fact, the defined pull events will become the milestones during the execution phase. Note that at this moment the focus is on completing the VFD, by filling its Concurrent Engineering and Flow Definition Sub-matrices.

12.2.1 Milestone Chart and Progress Board

Both the milestone chart and the progress board must be updated and reviewed at each team meeting. Activities will progress from Not Checked Out ≫ Checked Out ≫ Done through the progress board, and the team shall control the work pace according to this progress, also considering any risk mitigation and issue-solving activities added to the backlog.

12.2.2 Risks and Issues

Advancing through the PDP is like climbing a ladder, after each step seeing further ahead. As a consequence, more knowledge about the development challenge is built and better understanding of the related risks is obtained.

During the planning activities, you might find new risks and redefine previous ones, particularly when related to the team structure and capacity and aspects that might impact on the activity timely execution (i.e., aspects related to suppliers). Whenever a risk or issue is identified, mitigated, or solved, this filed is revisited and updated.

12.2.3 Fill the VFD's Concurrent Engineering and Flow Definition Sub-Matrices

At this moment, you have already created the value proposition (PD program scope), and now you must determine what functional divisions will need to participate during the development execution, what they will do (which value delivery function they will work on), and when they will do it (the pull events' scope).

The following VFD filling occurs according to the steps presented in Chap. 9.

12.2.3.1 Identify the Value Delivery Teams

The lean PD flow is achieved by product teams with all necessary skills to drive the general design, detailed engineering, prototyping, testing, procurement, equipment and production planning activities with no/minimum waste.

Integrated product teams, when effectively implemented (as seen in Chap. 2), greatly improve the use of human capital during PD and help provide a better understanding and communication among the various stakeholders.

The full development team encompasses all and only the necessary people to develop the alternatives chosen to be carried out during the development project, and to deliver the complete value items set through the designed functional architecture.

You shall consider "value delivered teams" subsets of people from the company's functional areas related to the PD program's product/service. These functional areas are not limited to engineering and production, but also might include the complete value chain areas (acquisitions, marketing, services, etc.) This is particularly true when the results from the development projects include not only (if any) product, but also services or even new value stream.

Note that the set of teams is bounded by the LPDO's organizational structure and by the results from the make or buy analysis which defines what is going to be acquired/supplied and what is going to be developed internally. Therefore, different LPDO might develop the same product/service using diverse value delivered team sets.

The stall recovering system example, for instance, included both value delivered teams related to the product and to the company's value chain which includes the suppliers:

- MEC—Mechanical Engineering
- ELE—Electronic Engineering
- SYS—Systems Engineering
- CHE—Chemical Engineering
- AER—Aeronautical Engineering
- QUA—Quality Engineering
- IND—Industrial Engineering
- PRO—Production Department
- CLI—The Client himself
- CE—Chief Engineer
- HOM—Homologation Department
- LOG—Logistics Department
- ADM—Administrative Department
- SPP—Suppliers of Pieces and Parts
- SLT—Suppliers of Labs and Test Facilities

12.2.3.2 Define the Contributing Roles of Each Value Delivery Team

Each team's role in contributing to deliver a particular function shall be mapped in the VFD's Concurrent Engineering Sub-matrix. During this mapping we recommend using the Role and Responsibility Charting (RACI) notation, as presented in Table 12.2.

By setting this relationship among functions and teams, the need of concurrent engineering becomes evident; the VFD creates a visual for which teams need to communicate and cooperate in order to guarantee that a certain function will deliver all the related pulled value.

Considering the [Receive Pilot Command] function in Table 12.3, all the linked teams contribute somehow to developing it. Therefore, the PD success is only achieved if all the participant teams perform real concurrent engineering.

Table 12.2 RACI mapping

		Description	How many in this role?
R	Responsible	Work on options and consequences; makes recommendations; coordinate the remaining of the group	Usually one, but sometimes more $(1 - n)$
A	Approver	Makes the decision	One (1)
C	Consulted	Makes recommendations	Varies from none to many $(0 - n)$
I	Informed	Get informed of the decision after it is made	Varies from none to many $(0 - n)$

Table 12.3 Concurrent engineering sub-matrix

MEC	R	R	R	C	R	R	R	R
ELE	R	R	–	–	R	R	R	R
SYS	R	R	C	C	R	R	R	R
CHE	–	–	–	R	–	–	–	R
AER	–	–	R	C	R	R	–	–
QUA	C	C	C	C	C	C	C	C
IND	C	C	C	C	C	C	C	C
PRO	C	C	C	C	C	C	C	C
CLI	C	C	C	C	C	C	C	C
CE	A	A	A	A	A	A	A	A
HOM	A	A	A	A	A	A	A	A
LOG	I	I	C	C	I	I	I	C
ADM	I	I	I	I	I	I	I	I
SPP	C	C	C	C	C	C	C	C
SLT	C	C	C	C	C	C	C	C
	Receive pilot command	Receive test engineer command	Create air resistance	Launch parachute	Jettison trailing cone	Jettison parachute	Lock the system	Execute ground test

12.2.3.3 Define Preliminary Pull Events

The challenge to schedule within a complex (and even multi-project) environment is to schedule in only the details that accomplish the objectives—avoiding the waste of excessive information and false sense of control. Intermediate dates are crucial to manage limited resources across multiple programs and these dates have to be approached with rigor and precision.

The pull events are the backbone of the value flow and are important moments to knowledge capture; by pulling the value delivery, they allow the planning to reach execution. Every pull event is associated with physical progress evidences (e.g., models, prototypes, start of production, etc.).

As a result, the company aligns and tiers engineering cadence with lower-level events designed to support higher level program events. Engineers and suppliers come to these reviews with prototypes, test results, open issues, and so forth, so that the CE can determine (at the source) whether the program is where it is supposed to be. Later in the process, the CE schedule physical prototype builds and part coordination events to the same effect. Engineering leaders meet periodically with the CE to review the program status, open issues, and performance metrics, which are posted on the PDVMB.

In this context, "boundary objects" (models, prototypes, tools, and activities that allow the sharing of knowledge and information across the organization and/ or areas of knowledge) facilitate integration, providing a common reference for the team [2, 3].

To define a sequence of preliminary pull events, the development team can use the enterprise's standard process (if there is one), reuse historical information from previous projects, or consider best practices from the industry. For the stall recovery system example, twelve pull events were preliminarily defined, as adapted from the company's standard development process (Fig. 12.3). Once two or more events have activities occurring in parallel, they cannot be characterized as phase gates.

Defining pull events is a tricky job. We recommend that you have what we call a "stairway building mindset," where you design the steps leading you to the PD vision. At each step you should have gained more knowledge and become more confident with the PD success. We also do not expect that you set a pull event that

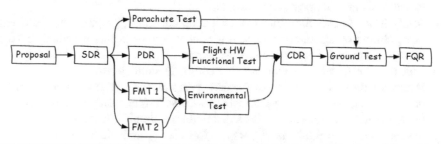

Event	Objective
Proposal	To have a feasible technical proposal approved
System Design Review (SDR)	To have the systemic conception of the product approved
Preliminary Design Review (PDR)	To have the subsystem's preliminary engineering designs approved
Functional Model Test 1.1 (FMT 1.1)	To have the riser cutter functional model approved
Functional Model Test 1.2 (FMT 1.2)	To have the trailing cone cutter functional model approved
Functional Model Test 1.3 (FMT 1.3)	To have the riser lock functional model approved
Functional Model Test 2 (FMT 2)	To have the mortar functional model approved
Parachute Test (PT)	To have the parachute functional model approved
Flight Hardware Functional Test (FHFT)	To have the pilots and flight engineer panel functional model approved
Environmental Test (ET)	To have the environmental model of the integrated system approved
Critical Design Review (CDR)	To have the detailed design approved
Ground Test (GT)	To have the system operationally approved
Final Qualification Review (FQR)	To have the system qualified by the client

Fig. 12.3 Pull events

gives you low confidence results and that requires you to reconfirm aspects that were on its scope: you should keep walking, but not in circles.

12.2.3.4 Relate the Pull Events to the Value Items and Risks

A pull event must be related to at least one value item and/or risk, and each value item must be checked by at least one pull event. A pull event's scope is defined by the set of related value items and risks, according to the MoE verifications that would be executed on these exact value items [2, 3]:

- **Inspection**: An action of observation, visual examination, or investigation against relevant documentation to confirm the compliance of the material or system with the technical requirements.
- **Analysis**: A check action through evaluation equations, graphs, data reduction, extrapolation of results, or reasoned technical argument, that specified requirements for a material or service have been met.
- **Calculus**: Performing mathematical or computer simulations.
- **Demonstration**: The display of features, performance, and operational capacity of an item, equipment, or system where success is found only through behavioral observation and/or results. Tests that require a simple quantitative verification measure, such as weight, size, time to perform tasks, are included in this category.
- **Test**: Verification of action, through the full exercise of the item, equipment, or system under appropriate controlled conditions, in accordance with approved test procedures. The test can be subsystem (T1) and the integrated product (T2).

Table 12.4 shows how the [realign the aircraft] value items and the risks were related to the defined pull events (Fig. 12.3).

12.2.3.5 Refine the Pull Event Set

Be checking the completely filled Flow Definition Sub-matrix, you can visually check if the scope of the pull events set that you planned makes sense. Having highly valued items less verified than low value items, and failing to check risk mitigation are some examples of common mistakes. Therefore, the preliminary pull event set shall be refined until it meets the following criteria:

(1) Is the set capable of verifying the progress of the effective value incorporation and the delivery of the project execution?
(2) Is the set balanced according to the value item's importance, where it is rare to expect less relevant value items being tested more thoroughly than the more relevant ones?
(3) Does the set represent the value flow in order to guarantee the information is pulled, not pushed?
(4) Does the set show the elimination of the risks that led to the development of multiple alternatives, allowing the combination and the reduction of the number of alternatives during the SBCE?

Table 12.4 Example of pull event's scope

Value	Value Item	Importance	Proposal	SDR	PDR	EMF1.1	EMF1.2	EMF1.3	EMF2	PT	FHFT	ET	CDR	GT	FQR
1 Realign the aircraft	1.1 Quick response to triggering	360	–	A	T	–	T1	–	T1	–	–	T2	A	T2	A
	1.2 Return to normal flight attitude	360	A	A	A	–	–	T1	–	T1	–	–	A	D	A
	1.3 Eliminate aerodynamical effects on the aircraft after use	108	–	A	A	T1	–	–	–	–	–	T2	A	T2	A
	1.4 Eliminate electrical effects on the aircraft after use	90	–	A	A	–	–	–	–	–	D	T2	A	T2	A
	...	...	...	...	...	...	...	...	...	...	...	...	...	...	...
Identified risks	R.1 Parachute does not fit into the mortar	12	–	A	A	–	–	–	,	D	–	–	A	D	A
	R.2 Delays on parachute delivery	12	–	–	–	–	–	–	–	A	–	–	–	A	–
	R.3 Not jointly achieve the release speed and the maximum retreat strength specifications	16	–	–	A	–	–	–	T1	–	–	T2	A	–	A

(continued)

Table 12.4 (continued)

Value	Value Item	Importance	Proposal	SDR	PDR	EMF1.1	EMF1.2	EMF1.3	EMF2	PT	FHFT	ET	CDR	GT	FQR
	R.4 Do not master the gas generator technology	12	–	–	A	–	–	–	T1	–	–	T2	A	D	A
	R.5 Requirement change requests from client	12	A	A	A	–	–	–	–	A	A	–	A	–	A
	R.6 Not be able to buy the mortar MP in time	12	A	–	–	–	–	–	–	–	–	–	–	A	–
	R.7 Pyrotechnical components validity does not meet the lifetime and/ or maintenance requirements	12	–	–	–	–	–	–	–	–	–	–	A	–	A

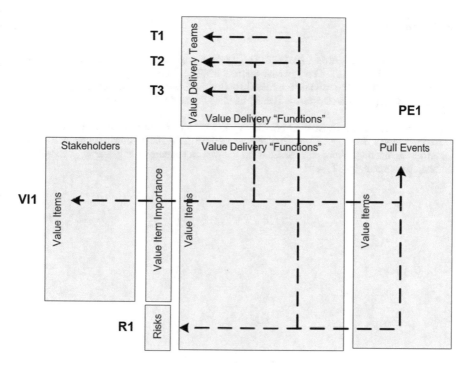

Fig. 12.4 The relation among the VFD's core elements

  12.3   A Practical View

At this moment, after having identified the value items set, and defined the SBCE strategy, the value delivery team, and the pull events set, the complete VFD shall be checked and balanced against the development project cost restrictions. It must be done before starting the execution phase.

Even though they reduce risk, there is a cost impact for both the SBCE, which includes more people to work on the multiple alternatives, and the pull events. Since they are waste mitigation strategies, they might increase the planned costs while reducing the likelihood of waste occurrence. Remember that if the wastes did occur, the expenditures would be even higher, but there is always the chance they will not happen. This is the dilemma of acquiring or not acquiring insurance.

Looking at the complete VFD (Fig. 12.4), the pull events are used to check that the actual implementation of the value delivery functions, which are made/built by the value delivery teams applying concurrent engineering, are indeed delivering all the value related to them (functions), while mitigating the associated risks.

While doing the final and complete VFD check, you should confirm that the pull event set give you confidence in delivering the value items (particularly the most important ones) while mitigating the risks.

References

1. Project Management Institute, PMI (2013) A guide to project management body of knowledge (PMBOK® guide), 5th edn. Project Management Institute, Newton Square
2. Pessôa MVP (2006) Proposta de um método para planejamento de desenvolvimento enxuto de produtos de engenharia (Doctorate Thesis) Instituto Tecnológico de Aeronáutica, São José dos Campos
3. Pessôa MVP, Loureiro G, Alves JM (2006) A value creation planning method to complex engineering product development In: Proceedings of 13th ISPE international conference on concurrent engineering, 2006, antibes. Leading the web in concurrent engineering, vol 143. IOS Press, Amsterdam, pp 871–881

Chapter 13
Execution Phase

By finishing the study phase, the development strategy is set and the LPDO is ready to start detailed engineering, prototyping, and production tooling. During the design and development activities (Fig. 13.1), all the module development teams should produce their deliverables in a fast and synchronized way, according to the sequence of defined pull events. This chapter uses the stall recovery system project example to present a stepwise execution of this phase's activities, where special emphasis is given on how to apply the SBCE and on how to identify and use the integrative design variables from the project's value set.

13.1 Introduction

During the execution phase the product envisioned during the study phase is physically developed and produced. Pull events foster concurrent engineering and sustain the sequence of rapid learning cycles which function as PDCA cycles; thus, making visible any quality problems and supporting knowledge creation. In this context, the execution activities' planning is decentralized, allowing different groups to realize their own plans to achieve the pull events.

We divided the execution phase in two sets of activities: (1) design and development activities; and (2) production/ramp-up activities. In the case of one of a kind and very personalized products production/ramp-up can be considered as part of the design and development activities, once only one product sample is going to be produced.

13.2 Using the Board to Guide the Design and Development Activities

In order to make the information visible and keep the development team synchronized, a PDVMB is also maintained during the Execution Phase. The greatest difference from the PDVMB used during the Study Phase is that the VFD is no

© Springer International Publishing AG 2017
M.V.P. Pessôa and L.G. Trabasso, *The Lean Product Design
and Development Journey*, DOI 10.1007/978-3-319-46792-4_13

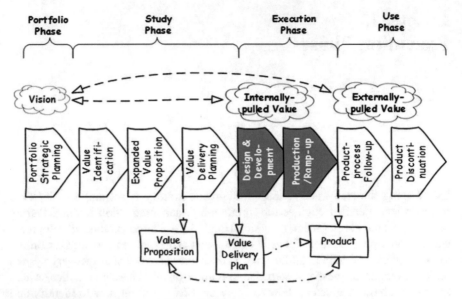

Fig. 13.1   Value proposition activities in the PDP

longer at the centre of the board; now the VFD is a reference of what has been planned, and the current state of the product being developed gets the focus.

The PDVMB filling sequence described in sequence (Fig. 13.2) will guide you during the Production Phase activities.

13.2.1   Setting Project Team

The project team directory must be updated in order to include the key people which will be responsible for actually executing the activities pulled by the several pull events.

In the case of small project groups, you can include the names of all the team members. On larger projects, you would rather include the point of contact from each participating group/area/supplier. Whoever you list here, this person must have the authority to decide and be accountable for his/her decisions.

13.2.2   Keeping the VFD Up-to-Date

Also in the VFD, the value delivered teams must be updated to match the actual project team. You do not need to put the team members' names here, but you must be able to map the value delivery team to the actual members, and vice versa. All the teams on the VFD shall have at least one person listed on the project team directory.

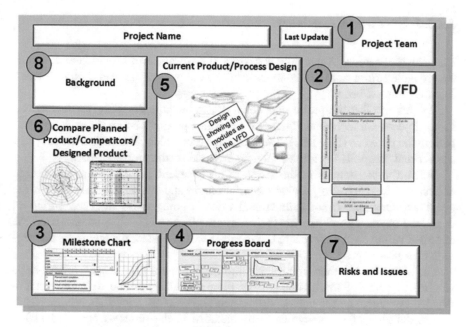

Fig. 13.2 PD visual management board filling sequence

As with any plan, we do not expect that everything will happen exactly as initially set in the VFD. During the execution, the team might find new risks, the value initially pulled might have some changes, new conflicts among value items might become apparent, you might decide to change some pull event's scope, and so forth. As a consequence, the team must always keep the VFD up-to-date.

13.2.3 Navigating Through Milestones

The milestone chart gives a program-level view of the design and development planned work, thus, supporting the meetings at the *obeya*. Each individual team shall have more detailed planning, even using bar charts, to support the execution and control of their work. We'd rather use a milestone chart at the program level in order to reduce wishful thinking which is not the case at the team level since they have much more knowledge to detail their activities and with low wishful thinking.

The minimum set of milestones should include the dates the team expects to have all the previously defined pull events executed. Other milestones, such as expected receiving dates from suppliers, should also be added here. This field is reviewed and updated during each team meeting.

13.2.4 Execution Kanbans

The activities which produce the necessary and sufficient information and materials are pulled from each development team by the pull events, and should be filled into the progress board.

Considering that the pull events network can be quite complex, particularly on large projects, we recommend keeping a separate progress board for each different path in the network (Fig. 13.3). Therefore, for each value item within the scope of a pull event, the pulled activities shall be included on the progress board.

When a value item is in the scope of a pull event, all the teams which help deliver this item have to provide the information and/or materials needed for the event. For example, if some functional value is going to be analyzed during an event, the teams should provide their designs showing how they incorporated the expected value; on the other hand, if the item will be tested, the teams should provide their prototypes for testing.

The activities are then controlled through the board until a new milestone is reached. If some of the activities from the previous milestone were not yet finished, they remain on the board, and the new activities are added.

The defining of the activities to be included in the progress board's backlog is done by answering questions such as: "What are the activities that each team related to the function Fx have to perform in order to deliver the value item Vy which is pulled for analysis by the event Ez?"

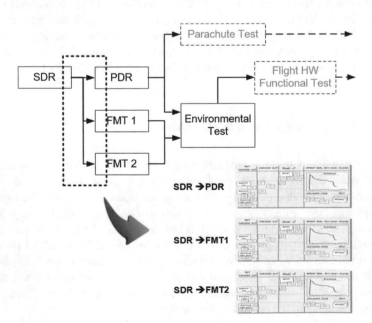

Fig. 13.3 Progress boards

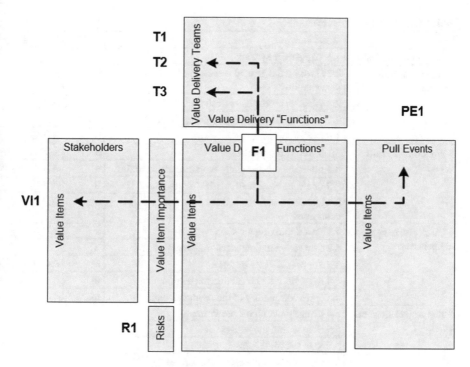

Fig. 13.4 Pulling activities from the teams

Figure 13.4 exemplifies the answering of this question where teams T1, T2, and T3, which are responsible for function F1, must perform the needed activities to deliver the necessary information or materials related to the value item VI1, according to pull event PE1's scope.

Considering the stall recovery system example, Table 13.1 lists the value items' subset that is related to the Parachute Launcher (PCL) development team and how they were included in the scope of the proposal pull event (the development activities are listed in parentheses in Table 13.2). Since this was the first development event, the only verification type used was analysis.

Table 13.2 lists the correspondent PCL development activities pulled by the proposal event. Whenever the method application suggested the use of concurrent engineering, the other participant teams are cited (when other teams are related to the same value item, the needed deliverables are pulled from all of them simultaneously). In the case of the development of multiple alternatives, the activities will be repeated for each alternative.

Table 13.1 Pulled activities example

Value	Value item	PCL	Proposal
1 Realign the aircraft	1.1 Trigger the system	a	
2 Safe and reliable operation	2.2 Have on the ground detection of system unavailability	a	A(1.1)
	2.2 Work when required (Reliability)	a	A(1.2)
	2.3 To not work when not required (Safety)	b	
	2.6 Useful life as…	b	A(1.3)
3 Work on aire rafts A and B	3.1 Mass no bigger than X	a	A(1.4)
	3.2 Interface mechanically with aircrafts A and B	a	A(1.5)
	3.3 Interface electrically with aircrafts A and B	b	A(1.6)
	3.5 Operate under the defined environmental conditions	a	A(1.7)
4 Quick and easy maintenance	4.1 Post deploy repair < X	a	A(1.8)
	4.2 Corrective maintenance time below T sec	a	
	4 3 Support Z years in stock	a	A(1.9)
	4.4 Must have technical documentation	m	
	4.5 Have traceability of the produced units	m	
5 The project must be viable	5.4 Comply with legal requirements	m	
	5.5 Stay within the budget	m	
	5.6 Stay within the deadline	a	
	5.7 Comply with the enterprise's rules	b	
7 Easy to manufacture and test	7.1 Adhere to the design for manufacturing and assembly guidelines	a	
	7.2 Have high rate of reuse of parts, processes, and technologies	m	
	7.3 Have complete and concise product, process and tests documentation	m	
	7.4 Have defined product and process acceptance criteria	a	
	7.6 Have more than one supplier to each procured item or raw material	m	

Table 13.2 Activities pulled from the PCL team

Proposal
(1.1) Determine alternatives "for the on-the-ground detection unavailability system" (TCJ, PCJ, TEQ)
(1.2) Include the PCL data in the system reliability estimate
(1.3) Include the PCL data in the useful life estimate
(1.4) Include the PCL data to the mortar mass estimate (PCH, TCJ, PCJ, LCK)
(1.5) Define the preliminary PCL mechanical interfaces
(1.6) Define the preliminary PCL electrical interfaces
(1.7) Estimate the PCL environmental condition limits (PCH, TCJ, PCJ, LCK)
(1.8) Estimate the time to post deploy repair
(1.9) Estimate maximum time to keep the system in stock

Fig. 13.5 Disciplines working together during the PD execution

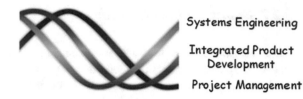

Systems Engineering

Integrated Product Development

Project Management

13.2.5 *Tracking the Current Product/Process Design*

Here, the product evolution through the execution of the planned design and development activities is presented, where the product will evolve from functional design, to detailed design, and then prototype.

This is the moment when system engineering, integrated product development and project management are going to walk hand-to-hand (Fig. 13.5). The VFD and SBCE should be applied in each of the system's engineering phases as the design and development evolves (see Chap. 9).

Whenever a prototype has been built, pictures from it shall be included if it's not feasible to have the prototype itself at the meetings. Particular caution has to be taken when moving from value items to detailed requirements; the correct traceability among them will guarantee that value alignment is kept. We strongly recommend the use of and requirements management software to keep track of this detailing and allocation and requirements allocation into products subsystems/modules.

The technical discussions among the team members will have the current product/process design or prototype as the main point of reference to solve issues. For instance, on the one hand, a person from maintenance would suggest that symmetry should be chosen for a given part geometry. On the other hand, a person from manufacturing would argue that asymmetry ought to be chosen for the same part in order to ease the setup of the lathe. Indeed, this discussion might go on with other people and technical areas. This is the true sign of interaction and concurrent engineering. The final version of the product is then defined by the development project leader, and shall accommodate in the best and balanced possible way the requirements of the whole product lifecycle.

13.2.6 *Mirror, Mirror on the Wall, Who Is the Fairest of Them All?*

During the whole execution phase, value is embedded into the product, and keeping track of this evolution, particularly when SBCE's alternatives are canceled, is an important indication of the development project's evolution.

We suggest keeping a radar chart where you can compare both the planned and actual product as well as the actual version of your competitors. This will also give you a good measurement for the business case of your upcoming product.

13.2.7 Bumps on the Miles Ahead

It is not uncommon that development teams, once they have started the execution, forget to keep making risk management; they tend to focus on the present, solving actual problems, and not acting proactively on identifying new risks if mitigating activities are not defined in the initial plan.

Risk management is an ongoing activity, whenever a risk or issue is identified, mitigated, or solved, this field is revisited and updated. You should invest some time during the periodical team meetings to try identifying which new risks are present and/or which previously identified risks have been surpassed.

13.2.8 Background and Change Management

Keeping the background field in the *obeya* helps with change management decisions. Whenever a new idea is brought into discussion, you must first check if it is compatible with the original product vision and then how it fits into the value items set.

If a new idea/pulled value/requirement is not compatible with the vision you either have to discard it or negotiate the changing of the vision.

13.3 Production and Ramp-Up Activities

During execution, all lean, quality, and design tools and techniques are applied to guarantee a *jidoka*-like and just-in-time-like development project (Fig. 13.6) [1]. By making information visible, it is easier to identify problems and waste. By stopping the chain of development events and investing time to understand the problems/waste root causes, you simultaneously guarantee solid/robust project deliverables and products and promote learning. By using techniques like 5S and Kanban, you both limit the work in process (WIP) and reduce the chance of waste occurrence.

The 5S is a quality technique that supports the organization and discipline at the workspace and reduces waste, where each "S" means:

- **Seiketsu** (清潔): Standardizing the best practices in the work area and maintaining everything in order and according to its standard.
- **Seiri** (整理): Organizing, making work easier by eliminating obstacles.
- **Seiso** (清掃): Keeping the work place clean and pleasing to work in.
- **Seiton** (整頓): Straighten, tidying up, arranging all necessary items so they can be easily selected for use.
- **Shitsuke** (躾): Training, sustaining and keeping discipline.

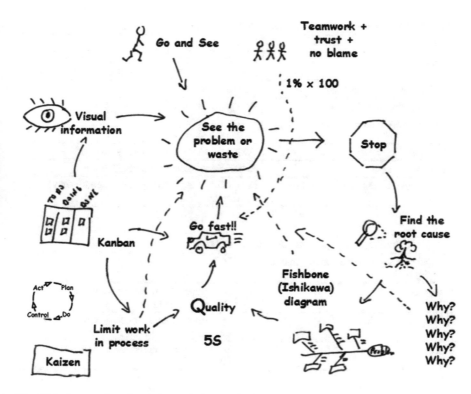

Fig. 13.6 Lean tools and techniques during production

 ⇨

13.4 A Practical View: Which DFX and/or DTX should I Use?

We suggest that you take advantage of the VFD in order to identify the need of DTX and/or DFX design tools. Since there is a fair amount of possible DTX and DFX techniques, it is not so easy to identify which of them would be useful during a particular project. In order to do that, we use the following rule of thumb.

For the DFX identification, look at the VFD's value identification matrix and ask, for each pulled value, if there is a related DFX. For DTX identification, look at the VFD's rework avoidance sub-matrix and check which value items are related to most, if not all, value delivery functions. These value items usually include an integrative design variable and are candidates to have an associated DTX.

Taking the stall recovery system example, Table 13.3 shows the high-level value pulled by the stakeholders and the related DFX. Similarly, Table 13.4 shows the value items associated to integrative design variables and their corresponding DTX tools.

Table 13.3 Stall recovery system's value items and DFX

Value	DFX
1 Realign the aircraft	The breadth of this value requires a deployment action prior to the DFX identification. See below.
2 Safe and reliable operation	Design for Safety
3 Work on aircrafts X and Y	Design for Modularity
4 Fast and easy maintenance	Design for Services Design for Modularity
5 Be viable	Design for Certification, Design to Cost
6 Provide learning	KBE—Knowledge Based Engineering
7 Be adequate to manufacturing and tests	Design for Manufacturing and Assembly and Design for Testing

Table 13.4 Stall recovery system's value items and DTX

Integrative Design Variable associated to Value items	DTX
1 Weight	Set a upper bound for the weight and use Design to Weight
2 Energy consumption	Set a upper bound for energy consumption and use Design to Net Power
3 Cost	Set a upper bound for cost and use Design to Cost

The value items deployed from the value [Realign the aircraft] are <Quick response to triggering>, <Return to normal flight attitudes>, <Eliminate aerodynamics effects on the aircraft after use> and < Eliminate electrical effects on the aircraft after use>. Each value item is associated to a system designed to deliver it. Find the system closet to a physical embodiment and apply the DFMA® tool to it. In this example, we chose the triggering mechanism to start with. The establishment of the minimum part count is always a good move to start the dialogue between the design and manufacturing technical areas.

Be aware of the design tradeoffs. Fear not! It's a rich experience for the product development team. The requirements for the several DFX techniques are— usually—in conflict to each other. The ideal product configuration for assembling might be the opposite for servicing the product. All people from the IPD design team should be committed to obtaining the best possible balanced results for the product, even if that means giving away some of his/her technical area expectations. The team coordinator has to assure that the final configuration of the product best balances all the lifecycle's requirements.

Still considering the stall recovery system example, the value items of having low weight, low energy consumption and low cost affect most of the value delivery functions, which indicates an opportunity to use DTX (Table 13.4); where, in this case, X means weight, net power and cost. Consequently, the functions should be balanced in these dimensions in order to optimize the product as a whole, which in this case can be achieved by the use of the Design to Weight (DTW), Design to Net Power (DTNP) and Design to Cost (DTC) directives.

In Chap. 2 we detailed the DTC technique, which can be used as template for DTW and DTNP. Note that your particular project may take benefit from using other integrative design variables, which were not mentioned in this book, but can be identified in the same way we presented here. You can then elaborate your own technical management cockpit (see Fig. 2.21) and place it at the Obeya. This cockpit will support you throughout your product design and development journey.

Reference

1. Blanchard BS, Fabrycky WJ (1988) Systems engineering and analysis, 3rd edn. Prentice Hall, New Jersey

Part V
On The Road

Part V discusses the reality when you put your lean development car on the road. Chapter 14 shows some of the bumps you might expect while on the track of your lean journey. Be assured, it is not a paved road!

Finally, Chapters 15 and 16 present two projects that applied this book's methodology. They aim to give you other perspectives to support you during your own lean journey (Fig. 1).

Fig. 1 The road leading to lean

Chapter 14
Project: Your Lean Journey

Following the understanding of all the Lean Wheel System's elements comes the moment to assemble your own wheel and start the lean journey by yourself. As we discussed in Chap. 7, adopting lean thinking in a company is a cultural change, as a consequence, changing a Product Development Process into a Lean Product Development Process means changing (sometimes subtly) the mindset as well as different aspects of your company.

In order to help you in this endeavor we are going to give you some advice on how to prepare yourself and how to proceed. But always keep in mind what we stated in this book's introductory chapter, since any company can copy techniques and practices or purchase the tools and technology used by any other company. Successful utilization of such techniques, practices, tools, and technology, though, depends on the ability to customize them in a way that makes them fit to the unique reality of the company using them. Remember that a lean tool can be used in non-lean way and vice versa.

14.1 Setting Your Attitude

As any athlete who makes a careful and thorough preparation before an event or competition, you must prepare yourself before starting your lean journey. This will help you avoid an over estimate of your expected gains, and will prevent you from getting discouraged if your progress is slow.

Every day, you should keep a *hansei* and *kaizen* attitude (humble and driven, never complacent) thus making your daily job a bit better each time. Also, work on becoming an example for your team, inspiring them to do the same.

By having the burden of responsibility on us, we tend to keep going even if something small happens. We postpone analyzing the problem and getting to a solution. These small problems are like sparks that will trigger lots of firefighting sometime in the future. A fire seldom starts big!

© Springer International Publishing AG 2017
M.V.P. Pessôa and L.G. Trabasso, *The Lean Product Design
and Development Journey*, DOI 10.1007/978-3-319-46792-4_14

Set specific, measurable, component level goals to build continuous improvement into each program. Celebrate every inch of your progress, it helps create a positive and winning mood in your team.

We also suggest that from time to time you double check your company against the product development low performance drivers we list in Appendix A. Appendix A is a good resource for you to find the presence of low performance drivers in your company, and for you to start working on them before they cause real problems.

We like to say that there are ten tenets for the lean leader:

1. Do understand the value that any initiative you take has to deliver.
2. Do make abundant use of go-and-see, being humble, and learning from everyone.
3. Do use what you learned to define a vision and make it a goal everybody will go towards.
4. Do create stable and steady improvement by taking one step towards the vision at a time; avoid great advances and regressions cycles (keep walking forward!).
5. Do think thoroughly before you act, and act fast after you think.
6. Do not be afraid of making mistakes; be afraid of not learning from them.
7. Be afraid of not making mistakes, maybe you are living an illusion.
8. Do accept your mistakes and let the team know that you will accept theirs while progressing towards the goal.
9. Do first, then say (be the reference).
10. Do not have a complacent attitude towards waste.

14.2 Before You Begin the Journey

Before you begin, you have to make sure you have all the gear and have assembled the right wheel for your journey.

First and foremost: make sure you have top management support. Any cultural change success is doomed if you have no support from high above. A top management public support acts like a force pulling the changes. Without it, any vision you define is your own and frail since it is not corporate aligned.

Remember that front loading the product development process means changing priority of how the time, money, and resources will be used (see Chap. 2). Therefore, how do you think you would be allowed to do so without upper support? For instance, how do you think you would be allowed to perform SBCE or to spend time during *kaizen* sessions?

Second, you must assemble a credible transformation team. The people responsible for driving lean must have active line involvement. Driving lean from a staff function with no active line roots reduces the initiative credibility [1].

Finally, create energy, critical mass, and awareness by communicating the change and educating the people. We are not talking about extensive training and materials rolled out to everyone before taking any action. We are talking about making the impacted group aware about what is going to happen and the expected benefits to be achieved.

Setting expectations is paramount. For instance, a manufacturing company in Brazil started implementing lean manufacturing in two of its sites at the same time. In one of the sites the initiative worked smoothly, the other went on strike. The reason for that: good/bad communications on what to expect from the changes. Even though this was a manufacturing process change and not a product development process change, it tackled the same issue we are talking about—changing culture and mindset.

14.3   Setting a Plan for Your Trip

Defining or changing your company's own development process is a tricky job. Some companies defined their process solely based on benchmark and on incorporating accepted best practices. Even though these sources give them good directions, you must cautiously plan your lean journey. We suggest taking the following steps:

1. Set the vision of your new product development process.
2. Identify the value through your product development process value chain: each company has a different value chain and you must understand what and who are you dealing with, how and which value they perceive.
3. Identify the possible wastes preventing the full value to be delivered: your process should be robust against waste. Considering the PDP peculiar particularities (Chap. 1), perform what-if analysis where you consider different scenarios and its possible issues and outcomes will help you identify possible waste causes. The comprehensive waste drivers list presented in Chap. 6 gives you a good start.
4. Identify the best practices, tools, and techniques (labeled and not labeled lean) that would help you to reduce/eliminate these wastes. Remember that the tools Toyota or any other successful company that applies the Lean Philosophy uses the tools that best fit them. It's not necessary for you to use exactly the same tools, but you shall pursue the same goal of internalizing the philosophy. Therefore, be creative about the tools while making sure that they have sufficient literature to support its correct use. By the way, using a new tool often is easier than using an old tool in a different way.
5. Reflect on your organization's maturity level in order to incorporate the alternatives of best practices, tools, and techniques you have chosen previously. Remember the climbing the stairs metaphor (Chap. 4), if your company is actually on the first step, jumping to the twentieth step might be very dangerous. Be careful and respect the limitations imposed by your people and process maturity level by taking care while climbing the stairway.
6. Plan the steps of your stairway (they will act as pull events).

Table 14.1 Using the VFD to support your lean journey

VFD element	Adaptation to supporting your lean journey
The product	What is your future PDP vision?
Stakeholders	Who is involved in performing, supporting, and receiving the results from your PDP?
Value items	What value do these stakeholders pull?
Measures of effectiveness	How is the value these stakeholders perceive present in the PDP?
Value delivery functions	Which are the PDP's activities?
Risks	What are the perceived risks to the improvement project? How can the company's maturity level affect the improvement process?
Value delivery team	Who performs the PDP activities? Who is going to support the improvement process project?
Pull events	What is the game plan for making the process changes, incorporating the best practices, tools and techniques, while mitigating the identified risks? What are the steps from my planned stairway?

As you could have imagined, the Lean Product Development Process we presented in Chaps. 9–13 can be easily adapted to support your PDP improvement (Table 14.1).

Be sure, though, that no plan gets away unchanged after execution. Indeed, the only assurance you have after you finish a plan is that it will not happen exactly that way! Therefore, you will need double the energy you use to plan it to execute it.

14.4 On the Road

While on the road, use the continuous improvement (Chap. 6) to create a stream of changes into the PD process in a way that the PD team can absorb and steadily grow by incorporating the best practices, tools and techniques.

The cross functional group composed of the value delivery teams should do the climbing of each step up your stairway of pull events as a PDCA cycle.

Different from physical products, where every piece, module, or system typically keeps its characteristics and functionalities after being developed, processes cannot be left alone. As we discussed in Chap. 6, a process, if left by itself, tends to erode and lose performance. Consequently, a sustaining team must support the process owners until they get used to the improved way and is capable of doing the continuous improvement by themselves.

Use the PDVMB to review and report progress and metrics. We emphasize the benefit of using a radar chart for showing the project progress on delivering value as each pull event is executed (Fig. 14.1).

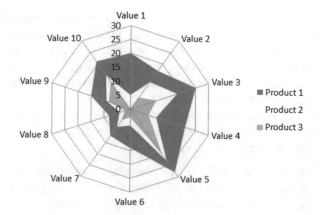

Fig. 14.1   Visually keeping track of the progress

Never stop going and seeing. Identify the improvement effect on the total PD system; pay special attention to identifying and addressing the issues of linkages and flow between processes.

14.5 Bumps on the Track

One thing we can assure you: the road ahead of you is neither paved nor smooth, and that your trip will be eventful. From our experience, the main sources of bumps you might expect while on the lean product development journey's track are related to:

1. Breaking the roots from the traditional paradigm;
2. Solving misunderstandings of the lean philosophy; and
3. Falling in the common pitfalls.

14.5.1 Breaking the Roots from the Traditional Paradigm

Even motivated teams struggle to give up on the actual paradigm. This is true if they are either in a comfort zone or facing a crisis (although when in a crisis people's minds are more open to change).

Plain and simple: people do not invest the time to really understand the new way; they often jump into biased conclusions. By really understanding we mean recognizing:

1. What remains the same as the previous way: When changing a complex process like the PDP, we expect that some things (maybe lots of them) remain the same. This is particularly true while applying continuous improvement.

2. What is similar to the previous way: This is the tricky part, when people jump to conclusions by (wishfully) thinking they understood it correctly. Since we measure new thing using our previous knowledge, if you leave the people alone to understand the new way, their previous experience will lead them to the wrong conclusions. This is where training, communication, and particularly the mentor-mentee system make a difference.
3. What is completely different: Even though this part is easy to recognize, here is where skepticism reigns. Again, training, communication, and the mentor-mentee are a must here.

Even though the traditional and the lean way aim to design and develop the best product/service, in reality they achieve different results. Companies applying the lean way deliver better products (better fit to the pulled value).

14.5.2 Solving Misunderstandings of the Lean Philosophy

Although being lean means focusing on delivering value while reducing/eliminating the waste, the first aspect that comes into people's minds when facing the "lean label" is "waste complete elimination." They expect doing more with less.

Some people even try to take advantage of the transitioning to lean effort by eliminating work they are not fond of doing, regardless of whether it delivers value or it is a necessary support to the value delivery activities.

For the people that believe lean = waste reduction, speeches advocating front loading the PDP and the use of SBCE are paradoxical and confusing. This is the reason that any change effort (transitioning to lean in particular) should be preceded by a well-designed and executed communication plan.

14.5.3 Falling in the Common Pitfalls

As previously listed in the ten tenets for the lean leader, the common pitfalls also have to be constantly in your mind. While the former say what you should do, the latter lists what to avoid doing.

The common pitfalls are:

- Aimlessly applying tools: Remember that it is not the tools that make you lean, but the philosophy behind them. Lean labeled tools can be applied in a non-lean way, and the other way around is also true. How do you know you are doing it right? Simple, the implementing of individual tools must create positive impact on the development process flow.
- Waiting to achieve perfect stability before getting started: It is true that a perfectly unstable process is harder to improve, but once you have a clear vision of where to go, you can start walking even if the process is not completely stable.

Waiting for perfect stability is like missing the vision for the perfect stability of the current process.

- Excessive communication without action: This is the same case as the story of the boy who cried wolf. Communication without action gives the sense of false start and erodes the initiative's credibility. We learn lean by doing, not by listening or reading.
- Making kaizen events an end onto themselves: Some companies define a kaizen quota for each sector to perform in a period. While these kaizens help teaching and incorporating the continuous improvement routine into the corporate culture, you should balance these learning kaizens with real improvement kaizens. By real improvement we mean those planned kaizens that have a system focus [1, 2].
- Complicating when you can keep it simple: The world is already getting more and more complex by itself, avoid adding more unnecessary complexity. Remember that this unnecessary complexity will only lead to waste.
- Giving responsibility to an underqualified/under experienced transformation team: Some companies send a few people to a training session and expect them to be experts, thus giving the responsibility to make a cultural change that they are not completely prepared for. [2] This ends up burning the credibility of the lean initiative.
- Letting outside experts do it for you: In order to understand the lean philosophy, one learns by doing. [1, 2] You can (and should) take the advantage of having external mentors, but they are the mentors, you are the executer.
- Not taking into account the actual company/team maturity: What we learn from the literature and from other companies' experiences is much more about their final result than the difficult path they had to follow in order to achieve it. You have to consider your team/company's maturity level and make sure your journey considers advancing it from level to level. Jumping levels usually creates greater assimilation difficulties and the related natural resistance to change.

   ## 14.6 A Practical View

From our experience, there are some aspects you might consider doing, which can improve the chances to have a better ride.

- Put up a block diagram showing the sequence from preparation to execution plus bumps.
- Use what we discussed about the VFD in order to plan your lean journey. Indeed, the vision you want to achieve is the arrival destination and the "product" from your lean transformation project.
- Use the PDVSB to make visual your roadmap.
- Use kaizen events to overcome all the bumps you find on the road as well as for training.

- Do not hesitate to stop on the side of the road and get some help from experienced mentors.
- Remember that the wheel hub elements (value, waste, and continuous improvement) are the core of everything else, therefore do not stick to lean-labeled tools, but creatively apply the available tools.
- Utilize a small pilot to achieve quick results. These results can be used as success cases that will help market the lean transformation initiative as a whole.
- Remember that a good plan helps but does not guarantee the journey's success: there are no rules, only exceptions!

References

1. Morgan JM, Liker JK (2006) The Toyota product development system. Productivity Press, New York
2. Rother M (2010) Toyota Kata: managing people for improvement, adaptiveness and superior results. McGraw-Hill, New York

Chapter 15
Thermo Baby Development Project

Priscila Malaguti Guerzoni (SENAI-MG)
Evandro Junio Gomes Lima (SENAI-MG)
Guilherme Augusto Mendes Pereira (SENAI-MG)
Marcus Vinicius Pereira Pessôa (ITA)
Luís Gonzaga Trabasso (ITA)

This chapter presents a piratical application of the book's method during the *Thermo Baby* development project, which was prepared as partial fulfillment of the requirements for the 2016 Lean Product Development Course we offered in the Mechanic Engineering Graduate Program at the Instituto Tecnológico de Aeronáutica (ITA). The team's task was to design and develop, in a six-month period, a product on the rehabilitation/assistive industry. They were required to use the VFD and PDVMB as presented in Chaps. 10–13.

15.1 Introduction

The Sudden Infant Death Syndrome (SIDS). SIDS, also known as cot death or crib death, is the sudden unexplained death of a child less than one year old. SIDS usually occurs during sleep, and the cause of death remains unexplained even after a thorough autopsy and detailed death scene investigation. There is usually no evidence of struggle and no noise produced.[1]

The motivation for this development is the fact that in Brazil, as in the rest of the world, SIDS is the main cause of healthy babies' death until they are one year old.[2]

---

[1]https://en.wikipedia.org/wiki/Sudden_infant_death_syndrome. Access in 11/16/2015.

[2]http://www.cdc.gov/nchs/data/nvsr/nvsr64/nvsr64_09.pdf. Access in 11/16/2015.

© Springer International Publishing AG 2017
M.V.P. Pessôa and L.G. Trabasso, *The Lean Product Design and Development Journey*, DOI 10.1007/978-3-319-46792-4_15

15.2 Background

The PD project vision defined by the team was to develop a device to monitor and help prevent SIDS. These devices work based on either: (1) heartbeat monitoring; (2) breath monitoring; or (3) body temperature monitoring. Considering the complexity and cost of the heartbeat monitoring and breath monitoring alternatives, the team chose to use the temperature monitoring approach. This decision did not compromise the application of the product development methodology, which was the exercise's main objective. As a result, the product vision was defined as:

The purpose of the *Thermo Baby* project is to develop a baby temperature monitoring device (named *Thermo Baby*) aimed at the Brazilian market that helps prevent Sudden Infant Death Syndrome (SIDS).

15.3 Current Condition

The *Thermo Baby* market is comprised of Brazilian mothers whose babies are one year old or less. Considering the initially estimated product cost range, the team focused on higher income groups from the targeted market.

According to Brazilian official data (IBGE 2013), every year about 3 million babies are born in Brazil and 32 % are born into high income families. The team also projected a 3 % market-share to be achieved by the end of the third year, which represents 28,800 products sold.

There are no direct *Thermo Baby* competitors being produced in Brazil, but we can find some substitutes:

- Regular heartbeat monitors (usually used at hospitals) which can be wired or wireless. These devices, however, are not proper for babies, since they cannot be fixed to them, and the baby's hand is too small, making it difficult to get a clear heartbeat signal.
- Heart rate monitor watches, although very precise and high tech, are originally designed for sports and are not available in babies' size.

Some direct competitors can be found as imports. For a benchmark, the group considered the following commercially available products (the names of the actual products are not presented here):

- Smart Baby Thermometer;
- Wireless Temperature Monitor; and
- Digital Thermometer with Bluetooth to Android.

By analyzing these competitors, the *Thermo Baby* price range should be less than US$100.00. For attaining this figure, we considered the price of the products themselves, shipping and handling to Brazil, and all the related import taxes.

15.4 Comparative Board

How the *Thermo Baby* compares to its competitors, based on the products' characteristics presented on their commercial data-sheets, is shown on the comparative
board (Fig. 15.1), where:

- **Comfort**: The product must be comfortable to the baby; therefore, it must not
  annoy him/her.
- **Remote monitoring**: It must not refrain the mobility of the mothers, nurses, or
  caretaker of the baby.
- **Safety**: The product must be certified as not harmful or hazardous to the baby.
- **Precision**: Temperature measures reliability.
- **Easy to install**: It can be installed by the person taking care of the baby without
  additional help.
- **Ready-to-use**: Once installed, the product is already functioning.
- **Data storage**: It allows the storage of historical data and the checking of tem-
  perature variation and possible deviations.
- **Easy maintenance**: Considering the engineering point of view and the target
  market need.

A rank ranging from 5 to 0 was defined based on the presence and/or intensity of
each relevant product characteristic as in Table 15.1.

Thermo Baby was planned to stand out from its competitors in terms of comfort, easy application (no belts or bands), and maintenance (easy-to-find replacement components in the Brazilian market). On the remote monitoring aspect, a

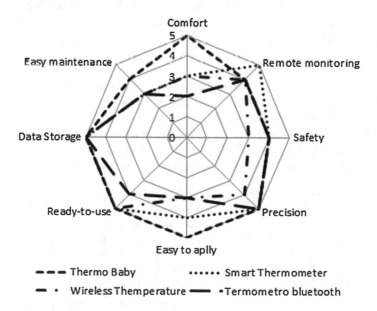

Fig. 15.1 Comparative board

Table 15.1 Ranking range

0	None	The value is not delivered at any rate
1	Insignificant	The value is delivered at a very low rate, mainly as a side effect from some other feature
2	Low	The value is perceived, but its rate does not fulfil the expectation from most any users
3	Medium	The value is delivered in the same level as the market average, thus not creating any differentiation to the product
4	High	The value is delivered at an above average level
5	Benchmark	This is the best product in class, in terms of delivering this particular value

product named *Smart Thermometer* has the best remote monitoring range, but the team considered a high ranking in this aspect as fulfilment of the actual market needs and that there was no need to outrank the competition.

15.5 Keeping Track of the Development Project

In order to set the development strategy and keep track of the development's progress, the team prepared a milestone chart and a progress board, as suggested in Chap. 10. In the planned milestone chart, as shown (Fig. 15.2), the team defined the week when each major group of activities should be finished. Considering the small size of the development team (4 people) and the short time available (4 months), the milestone chart was indeed very simple.

The Stakeholders' Analysis included the stakeholders' identification and prioritization. Once the stakeholders were identified, the Value Analysis included the value identification for each stakeholder (what they expect). It included interviews and "go and see" experiences to elicit the values that would guide the remaining design and development process.

Product design aimed to define the product's functional architecture which delivers the value as pulled by the stakeholders. It also included the analysis of different technological alternatives to develop the product's most critical parts and its related costs.

The Prototype phase involved the development of a proof of concept prototype. Proof of concept is documented evidence that a potential product or service can be successful. Finally, the prototype was validated by the selected focal group.

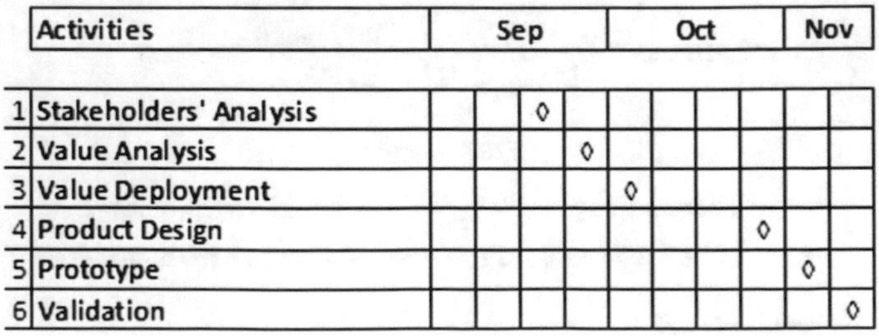

Activities		Sep		Oct		Nov
1 Stakeholders' Analysis		◊				
2 Value Analysis		◊				
3 Value Deployment			◊			
4 Product Design				◊		
5 Prototype					◊	
6 Validation						◊

Fig. 15.2 Milestone chart

By this brief description it is possible to realize how limited the time was between each milestone which created a real challenge for the development team.

In order to face this challenge, sets of activities between milestones were defined and executed. The team established a weekly meeting routine, in which a Progress Board was updated considering the work accomplished and the tasks still to be performed. At the beginning of every week, a set of activities to be performed was established and then each team member was defined as responsible for one or more activities. As long as the activities were executed, they were migrated from the "backlog" to "doing" and then "done."

15.6 Study Phase—Value Identification Activities

15.6.1 Stakeholders' Identification + Value Items Analysis

Through brainstorming, value chain analysis, and product life cycle analysis, the following groups of stakeholders were identified:

- **Pediatricians**, which might suggest that the caretakers buy the product.
- **Target Users**, which we divided into caretakers (parents, nurses, grandparents etc.) and the babies themselves. For simplification, we considered that the caretakers were also the buyers of the product.
- **Regulatory agencies** that define the certification requirements the product must fulfill in order to have its sales allowed.
- **Shareholders/company owners**, which will have financial benefit by selling the product.
- **The Development Team**, which is composed of the designers, production engineers, electronic engineers, mechanical engineers, and all the other specialties that will work during the design and development activities.
- **Sellers**, which are going to have the product in their store (either physical or virtual). Once we have competitor and substitute products in the market, it's important that we deliver value to them, in the sense they will find it easier to sell our product than the others.
- **Maintenance service providers**, which will make any necessary product maintenance and spare parts change.
- **Recyclers**, which will treat or process the product's parts so as to make suitable for reuse.
- **The Production Engineering**, which encompasses the people in charge of the product manufacturing and assembly activities.

The development team identified and ranked the stakeholders as in Table 15.2.

Figures 15.3 and 15.4 show how each stakeholder group pulls value through the product life cycle's phases. In order to identify these pulled values the team made interviews and performed "go and see" by visiting some recent mothers. The team eliminated the ambiguity from the initially pulled values, by detailing them into value items (Table 15.3).

Table 15.2 Stakeholder's prioritization

Primary	Caretakers, babies, pediatricians, shareholders, and regulatory agencies
Secondary	Production engineering and sellers
Tertiary	Development team, maintenance service, providers, and recyclers

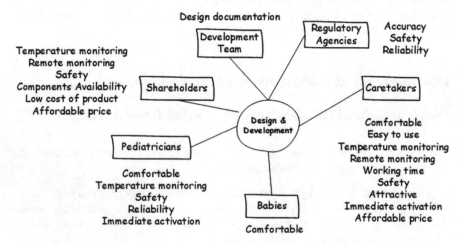

Fig. 15.3 Stakeholders and their pulled values at design and development stage

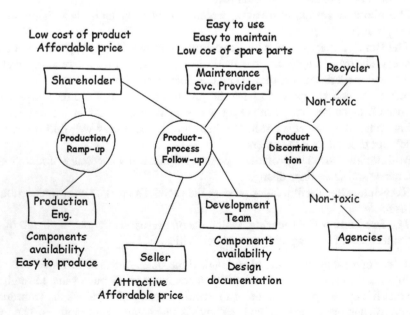

Fig. 15.4 Stakeholders and their pulled values at product ramp-up, product follow-up, and discontinuation stages

Table 15.3 Filling the VFD—from value to value items

Value		Value items	
1	Be comfortable to the baby	1.1	Be light on the baby's body
		1.2	Be small on the baby's body
		1.3	Provide soft touch to the baby
2	Easy to use by parent or caretaker	2.1	Be intuitive to apply
		2.2	Be easy to remove by parent or caretaker
		2.3	Have Intuitive commands
3	Provide temperature monitoring	3.1	Provide identifiable alert
		3.2	Log data for tendency analysis
		3.3	Have high reading accuracy
4	Allow caretaker's mobility	4.1	Allow remote monitoring
5	Work when needed	5.1	Have a proper battery life
		5.2	Have low downtime
6	Provide precise measure	6.1	Have precise sensors
7	Be safe for babies' use	7.1	Use non-toxic product
		7.2	Proper parts dimension
		7.3	Resist unwanted disassembly
		7.4	Hard to remove (baby)
8	Be reliable	8.1	Use high quality parts
		8.2	Use durable parts
		8.3	Babies' weight resistant
9	Be easy to maintain	9.1	Support several assembly-disassembly cycles
		9.2	High spare components availability
10	Attractive	10.1	Visually pleasant
		10.2	Good market communication
11	Ready-to-use	11.1	Immediate skin contact activation
		11.2	Few commands
12	Be profitable	12.1	Create intellectual property
		12.2	Low cost of product's parts
		12.3	Low cost of product's production
13	Be affordable	13.1	Fit on caretaker's budget
14	Be easy to build	14.1	Uncomplicated to produce/acquire
		14.2	Well documented product design

15.6.2 Prioritize the Value Items

Considering the stakeholders' prioritization and their ratings to each pulled value item, Table 15.4 shows the value item's final absolute and relative importance. During the remaining of the development, the development team kept considering the complete value items' set. The priority was used only when trade-offs were needed.

Table 15.4 Filling the VFD—value items prioritization

Value Item	Caretakers (Pri)	Babies (Pri)	Pediatricians (Pri)	Shareholder (Pri)	Regulatory Agencies (Pri)	Development Team (Ter)	Production (Sec)	Seller (Sec)	Maintain Svc. Provider (Ter)	Recycler (Ter)	Absolute Importance	Relative Importance
1.1 Be light on the baby's body	9	9	9	3				1			273	4,61%
1.2 Be small on the baby's body	9	9	9	3				1			273	4,61%
1.3 Provide soft touch to the baby	9	9	9	3	3			1			300	5,07%
2.1 Be intuitive to apply	9	3	3	3	1			1	1		175	2,96%
2.2 Be easy to remove by parent or caretaker	9	3	3	3	1			1	1		175	2,96%
2.3 Have Intuitive commands	3		3	2	1			3	3		93	1,57%
3.1 Provide identifiable alert	9		9	9	3			3			279	4,72%
3.2 Log data for tendency analysis	3		9	3	1			1			147	2,48%
3.3 Have high reading accuracy	9		9	9	9			3			333	5,63%
4.1 Allow remote monitoring	9		3	9	3			3			225	3,80%
5.1 Have a proper battery life	9		1	3	1			3		3	138	2,33%
5.2 Have low downtime	3			3	1			1			66	1,12%
6.1 Have precise sensors	1		9	1	3			3			135	2,28%
7.1 Use non-toxic product	9	9	9	9				3	1	9	343	5,80%
7.2 Proper parts dimension	9	9	3	9				1		3	276	4,66%
7.3 Resist unwanted disassembly	9	9	9	9				1	3	3	333	5,63%
7.4 Hard to remove (baby)	9	9	9	9				1			327	5,53%
8.1 Use high quality parts	3	9	3	9				1	3		222	3,75%
8.2 Use durable parts	1	1	1	3				3	3		66	1,12%
8.3 Babies' weight resistant	3	3	3	9				1	3		168	2,84%
9.1 Support several assembly-disassembly cycles	3	1	3		3	3		1	9	3	90	1,52%
9.2 High spare components availability	3	1	9		9	9		3	9	1	172	2,91%
10.1 Visually pleasant	9	3	3					9			162	2,74%
10.2 Good market communication	3	9	9					9			216	3,65%
11.1 Immediate skin contact activation	9	9	3	3				1			219	3,70%
11.2 Few commands	3	9	1		3			3	3		156	2,64%
12.1 Create intellectual property				9		3					84	1,42%
12.2 Low cost of product's parts				9					9		90	1,52%
12.3 Low cost of product's production				9							81	1,37%
13.1 Fit on caretaker's budget	9		1	9				9			198	3,35%
14.1 Uncomplicated to produce/acquire				3			9			3	57	0,96%
14.2 Well documented product design						9	9		9		45	0,76%

As a result, the most important value items were "use non-toxic product," "have high reading accuracy," "resist unwanted disassembly," "hard to remove (by the baby)," and "provide soft touch to the baby." These are values that guarantee the product reliable functionalities, safety, and comfort for the baby.

15.6.3 Define Measures of Effectiveness

After the value items' identification, each value item was associated with an identifiable and measureable effectiveness parameter (Table 15.5). The measures of

Table 15.5   Filling the VFD—measures of effectiveness

Value item		Measure of effectiveness	
1.1	Be light on the baby's body	1.1.1	<45 g
1.2	Be small on the baby's body	1.2.1	<25 cm^2
1.3	Provide soft touch to the baby	1.3.1	Focal group rate > 8
2.1	Be intuitive to apply	2.1.1	Focal group rate > 8
2.2	Be easy to remove by parent or caretaker	2.2.1	Pull-off test ISO 4624
2.3	Have intuitive commands	2.3.1	Focal group rate > 8
3.1	Provide identifiable alert	3.1.1	37.50 °C < Signal < 35.00 °C
3.2	Log data for tendency analysis	3.2.1	Data storage capacity: 120 registers
3.3	Have high reading accuracy	3.3.1	Receive the related certification
4.1	Allow remote monitoring	4.1.1	Range higher than 30 m
5.1	Have a proper battery life	5.1.1	Battery life > 48 h
5.2	Have low downtime	5.1.2	Charging time < 3 h
6.1	Have precise sensors	6.1.1	Measure in 0.01 °C steps
7.1	Use non-toxic product	7.1.1	Material toxicity = 0
7.2	Proper parts dimension	7.1.2	Parts > 3 cm^2
7.3	Resist unwanted disassembly	7.1.3	Apply DFMA directives
7.4	Hard to remove (baby)	7.4.1	Pull-off test ISO 4624
8.1	Use high quality parts	8.1.1	Deviation < 0.5 °C
8.2	Use durable parts	8.2.1	2 years guarantee os parts
8.3	Babies' weight resistant	8.3.1	Load resistance ⇒ 15 kg
9.1	Support several assembly-disassembly cycles	9.1.1	Apply DFMA directives
9.2	High spare components availability	9.2.1	Components available in the market
10.1	Visually pleasant	10.1.1	Focal group rate > 8
10.2	Good market communication	10.2.1	Focal group rate > 8
11.1	Immediate skin contact activation	11.1.1	First temperature measurement time < 5 s
11.2	Few commands	11.2.1	Number of commands < 3
12.1	Create intellectual property	12.1.1	A least one patent registered
12.2	Low cost of product's parts	12.2.1	Parts + production cost < US$50
12.3	Low cost of product's production	12.3.1	Parts + production cost < US$50
13.1	Fit on caretaker's budget	13.1.1	Product price < US$100
14.1	Uncomplicated to produce/acquire	14.1.1	Apply DFMA and DbF directives
14.2	Well documented product design	14.1.1	Documented product design

Table 15.6 Conflicting value items

Primary value		Secondary value
Hard to remove (baby)	YET	Be easy to remove by parent or caretaker

effectiveness allow the identification of the value presence at any moment during the project and are parameters against which the product should be verified during the pull events.

15.6.4 Identify Conflicting Value Items

Some value items may not be optimally delivered simultaneously. While the device must be hard to be removed by the baby, it must be easy to remove by the mother. Thinking about safety, the difficulty of being removed was considered the primary value and the ease of removal was considered the secondary value (Table 15.6).

In the *Thermo Baby* project, the team considered an adequate pull-off force to be 5 kgf/lb^2, which represents about 1/3 of the force of professional fixing tapes.

15.7 Study Phase—Value Proposition Activities

15.7.1 New Product Design: Functional Architecture

Considering the identified value set, the team designed a functional architecture capable of delivering it. Functional analysis was applied (see Chap. 11) resulting in the functions presented in Fig. 15.5. Note that it included functions related not only to the product operation, but also to fixing it to the base, and charging the batteries.

15.7.2 Risks and Issues

In order to deliver the value items through the designed functional architecture, the team identified and analyzed the likelihood and impact (Fig. 15.6) of four risks they decided to address during the development execution:

Risk 1.   Uncertainty regarding the material to be used for fixing
Risk 2.   Changes in legislation concerning emission's parameters. The team identified the possibility that the Telecommunication Agency would make changes in the radiation parameters which would affect the development project;
Risk 3.   Own technological limitations that prevent the board miniaturization; and

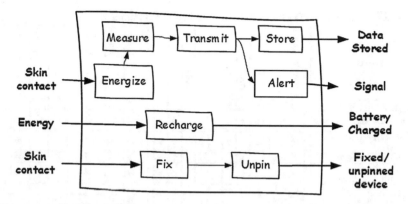

Fig. 15.5 *Thermo Baby* functional architecture

Risk	Likelihood (1-5)	Impact (1- 5)	Total	Grade
1	1	5	5	Low
2	2	5	10	Medium
3	1	3	3	Low
4	3	4	12	Medium

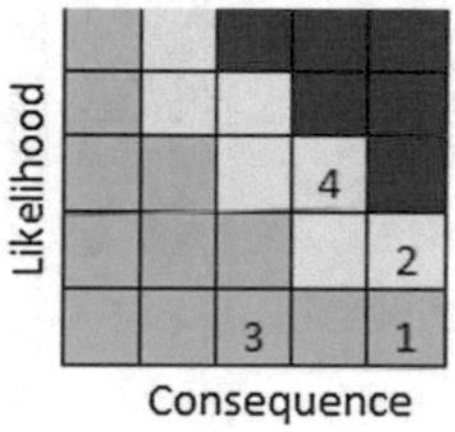

Fig. 15.6 Identified risks' likelihood and impact

Risk 4.   The launch of national competitor products. As some researches are of course in Brazil, it is very important to launch the product before its competitors, so they can get a new entrance advantage.

15.7.3 Fill the VFD's Value Rework Avoidance Sub-matrix

The functions identified during the functional analysis were transported to the VFD and related to the value items they support delivering. On Table 15.7 the column "Absolute Importance" is the total importance from each value item, as in Table 15.4; the column "Share" is calculated by dividing the absolute importance by the sum of the support from each function. Taking the value item "1.1 Be light on the baby's body," for instance, the absolute importance (273) was divided by the sum of the weight from each related function $(9 + 9 + 9 + 0 + 0 + 0 + 9 + 9 = 45)$, which resulted in shares of 6.07. Finally, the absolute importance is distributed to each function according to its supporting contribution.

Besides the value items, the value delivery functions were also related to the identified development risks. From this linking, the team considered the results from the likelihood X impact analysis and used the same distributive procedure as the one for distributing shares of the value items' absolute importance.

Table 15.7 Filling the VFD—functions and value items relationship

	Value Item	Absolute Importance	Share	Measure	Transmit	Energize	Store	Alert	Recharge	Fix	Unpin	Measure	Transmit	Energize	Storage	Alert	Recharge	Fix	Unpin
1.1	Be light on the baby's body	273	6,07	9	9	9				9	9	55	55	55	-	-	-	55	55
1.2	Be small on the baby's body	273	8,27	9	9	9				3	3	74	74	74	-	-	-	25	25
1.3	Provide soft touch to the baby	300	10,34	1	1	9				9	9	10	10	93	-	-	-	93	93
2.1	Be intuitive to apply	175	8,75	1	1					9	9	9	9	-	-	-	-	79	79
2.2	Be easy to remove by parent or caretaker	175	8,75	1	1					9	9	9	9	-	-	-	-	79	79
2.3	Have Intuitive commands	93	93,00	1								93	-	-	-	-	-	-	-
3.1	Provide identifiable alert	279	18,60	3	3			9				56	56	-	-	167	-	-	-
3.2	Log data for tendency analysis	147	14,70		1		9					-	15	-	132	-	-	-	-
3.3	Have high reading accuracy	333	37,00	9								333	-	-	-	-	-	-	-
4.1	Allow remote monitoring	225			9			3				-	169	-	-	56	-	-	-
5.1	Have a proper battery life	138	5,11	9	9	9						46	46	46	-	-	-	-	-
5.2	Have low downtime	66			9			9				-	-	33	-	-	33	-	-
6.1	Have precise sensors	135	15,00	9								135	-	-	-	-	-	-	-
7.1	Use non-toxic product	343	9,53	3	3	9			3	9	9	29	29	86	-	-	29	86	86
7.2	Proper parts dimension	276	8,36	9	9	9				3	3	75	75	75	-	-	-	25	25
7.3	Resist unwanted disassembly	333	7,40	9	9	9				9	9	67	67	67	-	-	-	67	67
7.4	Hard to remove (baby)	327	36,33								9	-	-	-	-	-	-	-	327
8.1	Use high quality parts	222	4,63	9	9	9			3	9	9	42	42	42	-	-	14	42	42
8.2	Use durable parts	66	2,54	9	9	1			1	3	3	23	23	3	-	-	3	8	8
8.3	Babies' weight resistant	168	9,33	9	9							84	84	-	-	-	-	-	-
9.1	Support several assembly-disassembly cycles	90	3,33						9	9	9	-	-	-	-	-	30	30	30
9.2	High spare components availability	172	8,19						3	9	9	-	-	-	-	-	25	74	74
10.1	Visually pleasant	162	10,80	3	3	3				3	3	32	32	32	-	-	-	32	32
10.2	Good market communication	216	6,35	9	3	3	1	3	3	3	9	57	19	19	6	19	19	19	57
11.1	Immediate skin contact activation	219	9,13	9	3	3				9		82	27	27	-	-	-	82	-
11.2	Few commands	156	17,33					9				-	-	-	-	156	-	-	-
12.1	Create intellectual property	84	4,67							9	9	-	-	-	-	-	-	42	42
12.2	Low cost of product's parts	90	2,14	9	9	3			3	9	9	19	19	6	-	-	6	19	19
12.3	Low cost of product's production	81	2,45	9	3	3				9	9	22	7	7	-	-	-	22	22
13.1	Fit on caretaker's budget	198	3,30	9	9	9	3	3	9	9	9	30	30	30	10	10	30	30	30
14.1	Uncomplicated to produce/acquire	57	1,54	9	9	1				9	9	14	14	2	-	-	-	14	14
14.2	Well documented product design	45	0,98	9	9	3	3	3	1	9	9	9	9	3	3	3	1	9	9
	(L X I)																		
	Total Value (TV)											1.404	919	700	151	412	189	930	1.213
Risk 1		5		0	0	0	0	0	0	9	9	0,00	0,00	0,00	0,00	0,00	0,00	2,50	2,50
Risk 2		10		0	9	3	0	0	0	3	1	0,00	5,63	1,88	0,00	0,00	0,00	1,88	0,63
Risk 3		3		3	3	0	0	0	3	0	0	1,00	1,00	0,00	0,00	0,00	1,00	0,00	0,00
Risk 4		12		1	1	3	1	1	1	1	1	1,20	1,20	3,60	1,20	1,20	1,20	1,20	1,20
	Perceived Risk (PR)											2,20	7,83	5,48	1,20	1,20	2,20	5,58	4,33

Figure 15.7 shows the best candidate functions to SBCE. These functions are the ones that combine a greater total value (TV) and perceived risk (PR) to successfully deliver this value. These are the functions which the development team should seriously consider to carry out multiple alternatives to accomplish during the development project.

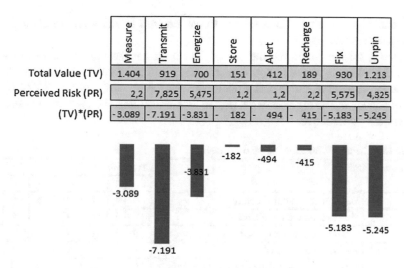

	Measure	Transmit	Energize	Store	Alert	Recharge	Fix	Unpin
Total Value (TV)	1.404	919	700	151	412	189	930	1.213
Perceived Risk (PR)	2,2	7,825	5,475	1,2	1,2	2,2	5,575	4,325
(TV)*(PR)	-3.089	-7.191	-3.831	- 182	- 494	- 415	-5.183	-5.245

Fig. 15.7 Best candidate functions of SBCE

There are three functions that are more critical. The first is the "transmit function" since there are legal discussions in Brazil about the allowed transmitting frequency that can interfere with the device configuration. The second and third are closely related, since they deal with "fixing" and "unpinning" the product, they are critical due to the amount of value they deliver and due to the uncertainty of the material that would be used as the fixing part in the device.

15.7.4 Finding Alternatives

The team's SBCE alternatives definition strategy consisted of a brainstorm where seven different ways of performing the critical functions were figured out (Table 15.8). The fixing and unpinning functions were considered together, since the chosen alternatives, when implemented, would do both.

In order to determine the alternatives that best fit the value set, the team made a Pugh analysis (positive [1], negative [−1], or neutral [0] correlation) verifying how each alternative behaved against the pulled value. Tables 15.9 and 15.10 show

Table 15.8 Seven ways to develop the product value delivery functions

	Transmit	Fix and Unpin
1	Wi-fi	Belt
2	Bluetooth	Adhesive
3	Radio	Brace
4	Infrared	Silicon
5	RFID	Sewn in
6	NFC	Velcro
7	Cable	Tie

the results for the alternatives related to the transmit function and to the fix/unpin functions, respectively.

Among all the solutions, the use of Bluetooth got the highest score to implement the transmit function, followed by Wi-Fi and infrared. In the case of the fix and union functions, the use of silicon received the highest score, followed by the use of adhesive. During this work, due to time constrains, the team developed only the winning alternative.

Table 15.9 Transmit function alternatives analysis

Value item		Transmit						
		1	2	3	4	5	6	7
1.1	Be light on the baby's body	0	0	0	0	0	0	0
1.2	Be small on the baby's body	0	0	0	0	0	0	0
1.3	Provide soft touch to the baby	0	0	0	0	0	0	0
2.1	Be intuitive to apply	1	1	1	1	1	1	−1
2.2	Be easy to remove by parent or caretaker	0	0	0	0	0	0	0
2.3	Have Intuitive commands	1	1	1	1	1	1	1
3.1	Provide identifiable alert	1	1	−1	1	1	1	1
3.2	Log data for tendency analysis	0	0	0	0	0	0	0
3.3	Have high reading accuracy	0	0	0	0	0	0	0
4.1	Allow remote monitoring	0	0	0	0	0	0	0
5.1	Have a proper battery life	0	0	0	0	0	0	0
5.2	Have low downtime	−1	1	−1	−1	−1	−1	−1
6.1	Have precise sensors	0	0	0	0	0	0	0
7.1	Use non-toxic product	0	0	0	0	0	0	0
7.2	Proper parts dimension	0	0	0	0	0	0	0
7.3	Resist unwanted disassembly	0	0	0	0	0	0	0
7.4	Hard to remove (baby)	0	0	0	0	0	0	0
8.1	Use high quality parts	0	0	0	0	0	0	0
8.2	Use durable parts	1	1	1	1	−1	1	1
8.3	Babies' weight resistant	0	0	0	0	0	0	0
9.1	Support several assembly-disassembly cycles	0	0	0	0	0	0	0
9.2	High spare—components availability	1	1	1	1	−1	−1	1
10.1	Visually pleasant	0	0	0	0	0	0	0
10.2	Good market communication	0	0	0	0	0	0	0
11.1	Immediate skin contact activation	0	0	0	0	0	0	0
11.2	Few commands	1	1	1	1	1	1	1
12.1	Create intellectual property	0	0	0	0	0	0	0
12.2	Low cost of product's parts	0	0	0	0	0	0	1
12.3	Low cost of product's production	0	0	0	0	0	0	0
13.1	Fit on caretaker's budget	0	0	0	0	0	0	0
14.1	Uncomplicated to produce/acquire	0	0	0	0	0	0	0
14.2	Well documented product design	0	0	0	0	0	0	0
	Total	5	7	3	5	1	3	4

Table 15.10   Fix/unpin function alternatives analysis

Value item		Fix/unpin						
		1	2	3	4	5	6	7
1.1	Be light on the baby's body	−1	1	1	1	−1	1	−1
1.2	Be small on the baby's body	−1	−1	1	1	1	1	1
1.3	Provide soft touch to the baby	−1	−1	−1	1	−1	−1	−1
2.1	Be intuitive to apply	−1	0	1	0	−1	0	0
2.2	Be easy to remove by parent or caretaker	−1	1	1	1	−1	1	1
2.3	Have intuitive commands	0	0	0	0	0	0	0
3.1	Provide identifiable alert	0	0	0	0	0	0	0
3.2	Log data for tendency analysis	0	0	0	0	0	0	0
3.3	Have high reading accuracy	0	0	0	0	0	0	0
4.1	Allow remote monitoring	0	0	0	0	0	0	0
5.1	Have a proper battery life	0	0	0	0	0	0	0
5.2	Have low downtime	0	0	0	0	0	0	0
6.1	Have precise sensors	0	0	0	0	0	0	0
7.1	Use non-toxic product	1	1	−1	1	1	1	−1
7.2	Proper parts dimension	1	1	−1	1	0	−1	1
7.3	Resist unwanted disassembly	0	0	0	0	0	0	0
7.4	Hard to remove (baby)	−1	1	−1	−1	1	−1	1
8.1	Use high quality parts	0	0	0	0	0	0	0
8.2	Use durable parts	1	1	−1	1	−1	−1	−1
8.3	Babies' weight resistant	0	0	0	0	0	0	0
9.1	Support several assembly-disassembly cycles	0	0	0	0	0	0	0
9.2	High spare components availability	0	0	0	0	0	0	0
10.1	Visually pleasant	−1	1	−1	1	1	−1	−1
10.2	Good market communication	0	0	0	0	0	0	0
11.1	Immediate skin contact activation	0	0	0	0	0	0	0
11.2	Few commands	0	0	0	0	0	0	0
12.1	Create intellectual property	0	0	0	0	0	0	0
12.2	Low cost of product's parts	0	0	0	0	0	0	0
12.3	Low cost of product's production	0	0	0	0	0	0	0
13.1	Fit on caretaker's budget	0	0	0	0	0	0	0
14.1	Uncomplicated to produce/acquire	0	0	0	0	0	0	0
14.2	Well documented product design	0	0	0	0	0	0	0
	Total	−4	3	−4	7	−1	−1	−1

15.8 Study Phase—Planning Activities

In order to better give priority and to avoid redundant activities, the development team members' responsibilities related to each value delivery function were defined and represented in an RACI chart at the top of the VFD (Table 15.11).

15.8.1 Fill the VFD's Flow Definition Sub-matrix

After identifying the value delivery teams, it's time to define the preliminary pull events which represent the backbone of the value flow. As pull events, the initially defined project milestones were kept and considered the objectives as presented in Table 15.12. The pull events were then deployed into the activities, which were put in the progress board's backlog for progress tracking.

Table 15.11 *Thermo Baby* RACI chart

	Measure	Transmit	Energize	Storage	Alert	Recharge	Fix	Unpin
Priscila	C	C	C	C	A	C	A	A
Guilherme	C	C	C	C	C	C	R	R
Evandro	R	R	R	R	R	R	C	C
Marcus	A	A	A	A	A	A	A	A
Luís	A	A	A	A	A	A	A	A

Table 15.12 Pull events' objectives

Event		Objectives
1	Stakeholders' analysis	To have the stakeholders' identification
2	Value analysis	To identify the stakeholders' value items
3	Value deployment	To have the measure of effectiveness of each value
4	Product design	To have the product preliminary designs
5	Prototype	To have the preliminary tangible product concept
6	Validation	To have the primary stakeholders' validation

15.9 Execution Phase—Design and Development Activities

During the execution phase, the sequence of pull events were performed. Considering the value to be delivered, the team applied some Integrated Product Development (IPD) techniques that best support this value delivering.

Table 15.13 shows the value pulled DFX, where:

- By using DFMA (Design for Manufacturing and Assembly), the team avoided different thickness in the product case, achieved a minimum number of parts by using multifunctional components, and avoided fixing elements (as rivet and bolts) (Fig. 15.8).
- By applying DFMt (Design for Maintainability) and DFSv (Design for Services), the team preferred easily found components, and designed a modular case where components with expected higher failure or replacement rates could be changed without affecting the whole system.
- By considering DFSft (Design for Safety), the team avoided sharp edges and designed in safety features to ensure that the product would not disassemble while in use.
- Finally, from DFRel (Design for Reliability), the team made sure that no failure mode would lead to misreading and to false positive and false negative measures; in this way, any system failure will shut it down, preventing false readings. Also, when choosing the components that could affect the measuring function reliability, the team preferred higher quality than lower price.

Table 15.13 Value items and related DFX

Value item		DFMA	DFM	DFSv	DFSft	DFR
1.3	Provide soft touch to the baby				X	
2.1	Be intuitive to apply				X	
2.2	Be easy to remove by parent or caretaker				X	
2.3	Have intuitive commands				X	
6.1	Have precise sensors					X
7.1	Use non-toxic product				X	
7.2	Proper parts dimension	X			X	
7.3	Refrain unwanted disassembly	X			X	X
7.4	Hard to remove (baby)				X	
8.1	Use high quality parts					X
8.2	Use durable parts					X
8.3	Babies' weight resistant				X	
9.1	Support several assembly-disassembly cycles	X	X	X		
9.2	High spare components availability		X	X		
11.2	Few commands				X	
12.2	Low cost of product's parts		X	X		
12.3	Low cost of product's production	X				
14.1	Uncomplicated to produce/acquire	X				

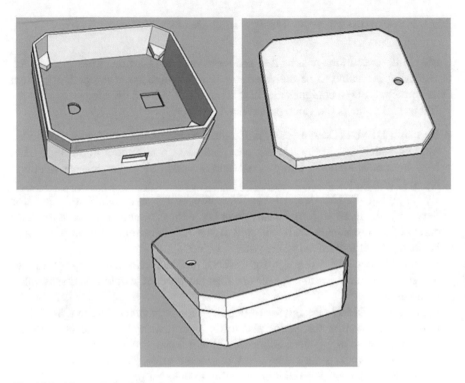

Fig. 15.8 *Thermo Baby* device features

Fig. 15.9 Proof of concept and final product concept comparative sizes

Although weight, size, robustness, and net power were identified as integrative variables from the value items set, they were not considered while developing the proof of concept prototype. They were considered in the market product concept, which is about three times smaller and half the weight of the prototype (Fig. 15.9).

15.10   Final Product Result

The *Thermo Baby* prototype was developed at the Senai Open Lab located in Belo Horizonte, Brazil. It took about 3 h of 3D printing to build the case that was tested and approved by the team engineers.

In order to deliver all the pulled value, software (a mobile phone app) that allows the communication between the *Thermo Baby* and a mobile phone was also developed. The device could then inform the temperature and alert in case of low temperature or overheating.

Figure 15.10 shows a software test before assembly, and Fig. 15.11 shows views of the prototype during the assembly.

The case is then enveloped by adhesive silicon that fixes to the skin and gives a very comfortable feeling. The product can be used more than 100 times without losing its adhesive characteristics, and has a low replacement cost. The product must be placed under the babies' clothes while sleeping.

The prototype cost was calculated considering the components' market price; considerable price reduction is expected in case of serial production. The prototype total cost was US$60:

- Case (3D printing)—US$10
- Components—US$30
- Technical hours—US$20

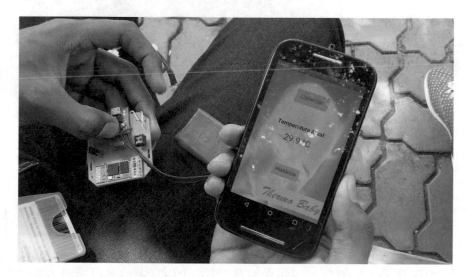

Fig. 15.10   *Thermo Baby* system test

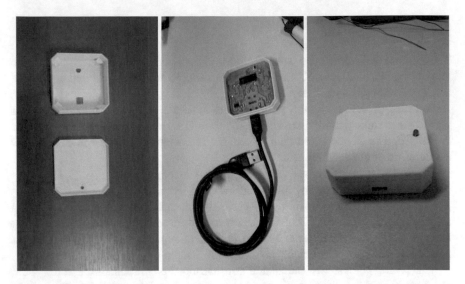

Fig. 15.11 The prototype's assembly

Fig. 15.12 *Thermo Baby* final concept

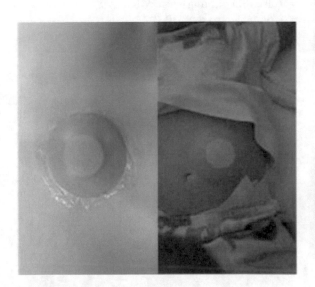

After the concept was proved and approved, the team studied similar accessible components that could be used in order to reduce the device's size. This resulted in a design half the prototype's size and three times thinner. This final product concept is shown in Fig. 15.12.

⇨ ## 15.11 A Practical View

By analyzing this example, we can perceive that the method gave sequence and priority to the development team. Even with limited time, they were capable to perform satisfactory work. Given more time, they might have achieved better results, but it is important to get the best results within the given market window.

We highlight that different teams or even the same team in different conditions would have different results for each of the method's steps. The real value from what we propose resides in the use of the method, rather than in the individual results themselves.

Chapter 16
Robot Based Flight Simulator Development Project

André Vinícius Santos Silva (ITA)
Wesley Rodrigues Oliveira (ITA)
Marcus Vinicius Pereira Pessôa (ITA)
Luís Gonzaga Trabasso (ITA)

This chapter presents the application of the lean product design and development method to the SIVOR (Simulador de Voo com Plataforma Robótica de Movimento—Robot Based Flight Simulator) project, which is a 3-year research project aiming to develop a flight simulator that combines the capabilities of a Full Flight Simulator (FFS) and of an Engineering Development System (EDS).

16.1 Introduction and Background

The certification campaign is one of the fundamental stages on an aircraft design and development process. Within this context, Flight Test Engineering is responsible for designing all the systems and procedures necessary to acquire and analyze flight test data. In general, these flights are done using either real aircrafts or simulation platforms. As flight testing using real aircrafts is more expansive and riskier than using flight simulators, the choice of using simulators is always preferred whenever it is possible.

In order to support the engineers during the aircraft development, an ideal simulation platform should both emulate a high fidelity cabin environment and be flexible to allow the pilot training and the optimization of the aircraft development process.

The SIVOR R&D Project is set upon this context. This project has been developed at Aeronautics Institute of Technology (ITA) in partnership with an Aircraft Manufacturing Company (AMC) Flight Test Team.

(*Purpose*) Develop a flight simulator with high fidelity environment and flexibility to enable the pilot training in the flight tests campaign and to assist on the aircraft design.

© Springer International Publishing AG 2017
M.V.P. Pessôa and L.G. Trabasso, *The Lean Product Design and Development Journey*, DOI 10.1007/978-3-319-46792-4_16

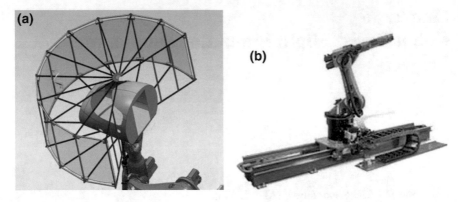

Fig. 16.1 Project constraints

(*Targets*) The specific targets are:

- Training pilots in a FFS (FAA—14 CFR Part 60) flight simulator for the execution of risky maneuvers;
- Optimize the development process of fly-by-wire control law during the aircraft design phase;
- Optimize the PDP cycle by leveraging the learning curve to the preliminary design phase.

As part of the scope of the defined research project, the simulator should include a robotic motion platform to representatively simulate the aircraft movement and the visual system should be attached to the cockpit (Fig. 16.1a). Besides the 6 degrees of freedom supported by the robot itself, the robot will move over a track, which gives the 7th degree of freedom (Fig. 16.1b). As a consequence, these aspects act as constraints to the development project, when choosing possible alternatives to perform the functions; while using the robot and the track is a must have condition, having the embedded visual system is a desirable condition, although it poses some challenges that are further discussed.

16.2 Current Condition

One can categorize flight simulators into training simulators and development simulators (EDS—Engineering Development Systems). While the former aims to train pilots to perform procedures and get acquainted to the aircraft they fly, the latter are engineering tools that support the aircraft development process.

The Full Flight Simulators (FFS) are the most advanced training simulators. FAA certifies FFS (FAA—14 CFR Part 60) ranging from level A to D:

- FAA FFS Level A—Requires a motion system with at least three degrees of freedom. This category is applicable only to airplane simulators.

Fig. 16.2 Stewart platform
(Hexapod)

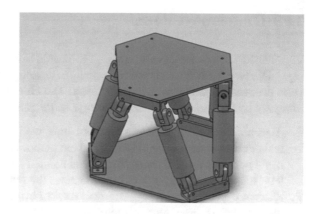

- FAA FFS Level B—Requires three axis motion and a higher-fidelity aerodynamic model than that of Level A. This category is the lowest level of helicopter flight simulator.
- FAA FFS Level C—Requires a motion platform with all six degrees of freedom, and a lower transport delay (latency) than those of levels A and B. The visual system must have an outside-world horizontal field of view of at least 75° for each pilot. It must also accurately replicate every aircraft system available from the cockpit.
- FAA FFS Level D—The highest level of FFS qualification currently available. It includes all requirements from Level C with additions. The motion platform must have all six degrees of freedom and the visual system must have an outside-world horizontal field of view of at least 150°, with a collimated (distant focus) display. Realistic instruments and sounds in the cockpit are required, as well as a number of special motion and visual effects.

The FFS training devices, though, do not have the flight envelope adapted to risky maneuvers tests and do not have the flexibility to change its cockpit environment and aerodynamics model. Although they fulfil the most demanding motion requirements, the Stewart Platform (Hexapod), over which they are mounted have limited dynamics and workspace, restricting the possible range of maneuvers (Fig. 16.2).

The EDSs are advanced engineering platforms, flexible and fitted to simulate and test different alternatives of the aircraft's systems, aerodynamics models as well as fly-by-wire control laws. They lack, though, the motion capabilities needed for training flight test pilots to perform risky maneuvers. They also do not need to fulfil any FAA/Part-60 requirements.

The SIVOR aims to both include representative motion and visual simulation capabilities, while keeping the flexibility of engineering platforms.

16.3 Comparative Board

To understand how the SIVOR concept compares to a FFS Level D and to an average EDS, five criteria were chosen, which the development team considers encompass the greatest challenges to the project. The chosen criteria were qualitatively analyzed by the design team, considering a score ranging from 0 to 30, where:

1. **Aircraft cockpit representativeness**: how well the simulator represents the real aircraft systems, the avionics and the audio. Grade 30 means that the pilot could perceive no difference from the simulated and the real cockpit, and grade 0 means that the cockpit does not resembles the real aircraft in any way:
2. **External environment representativeness**: how well the simulator represents the external environment; it includes factors such as: scenario detail richness, meteorological conditions, image resolution, 3D immersion and visual field, giving the pilot a closer feeling to the real flight. Grade 30 means that the pilot gets immersed into the simulation and grade 0 means that there is no visual system.
3. **Motion representativeness**: how well the simulator represents the aircraft's operation and maneuvers. Grade 30 means that the pilot can't differ from flying the real aircraft and flying the flight simulator, while grade 0 means that there is no motion system.
4. **Flexibility**: how easily new aircraft flight models, instruments and even entire cockpits can be changed, in order to support the engineers during the development of a new aircraft. Grade 30 means the designers can swap components and systems and grade 0 means that components and system are not replaceable.
5. **Cost**: comparative from the estimated development, operation and maintenance costs.

SIVOR was planned to stand out from its competition by providing a superior solution in terms of balancing motion representativeness and flexibility (Fig. 16.3). In order to achieve a FFS Level D certification, it could not fall behind its counterparts in the cockpit and visual representativeness criteria.

Fig. 16.3 Comparative radar chart

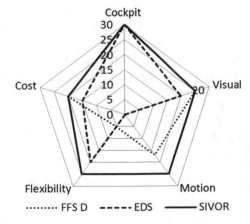

16.4 Keeping Track of the Development Project

As part of the research project funding application process, the team had to prepare upfront a detailed project schedule. As a consequence a milestone subset were picked out from the initial plan and used to keep track of the project. The milestones were divided into 6 modules: Robotic platform specification and acquisition (Module 1), Cockpit specification and acquisition (Module 2), Visual System specification and acquisition (Module 3), System integration specification and project (Module 4), Technical presentations of the project development (Workshops – Module 5), and Technical deliveries of important systems and equipment (Module 6).

The team also had weekly meetings when all the groups (including AMC engineers) presented its progress and issues and the plan for the following week was set.

Table 16.1 shows the milestone chart where:

Module 1, Milestone 1.1—Robotic motion platform technical specification ready: meaning that the specification is ready and reviewed, and thus the robot and track acquisition can be made. This is an early milestone due to the fact that the robot is build on demand, and the delivery time is 6 months after the order is placed.

Module 1, Milestone 1.2—Robotic motion platform requirements validation: verify the latency between the robot path calculator and the SIVOR external controller.

Table 16.1 Milestone chart

Module	Milestone	May-15	Nov-15	Dec-15	Jan-16	Feb-16	Mar-16	Apr-16	May-16	Jun-16	Jul-16	Aug-16	Sep-16	Oct-16	Nov-16	Dec-16	Jan-17	Feb-17	Mar-17	Apr-17	May-17	Jun-17	Jul-17	Aug-17	Sep-17	Oct-17	Nov-17	Dec-17
Mod 1	Milestone 1.1	▲																										
	Milestone 1.2								▲									▲										
Module 2	Milestone 2.1					▲																						
	Milestone 2.2						▲																					
	Milestone 2.3															▲												
	Milestone 2.4																				▲							
Mod 3	Milestone 3.1							▲																				
	Milestone 3.2																					▲						
Mod 4	Milestone 4.1										▲																	
	Milestone 4.2																							▲				
Módulo 5	Milestone 5.1	▲																										
	Milestone 5.2		▲																									
	Milestone 5.3									▲																		
	Milestone 5.4															▲												
	Milestone 5.5																						▲					
	Milestone 5.6																											▲
Módulo 6	Milestone 6.1							▲																				
	Milestone 6.2										▲																	
	Milestone 6.3																▲											

Module 2, Milestone 2.1—Cockpit configuration technical specification ready: the choices for flight instruments and flight commands should be defined, reviewed and approved.

Module 2, Milestone 2.2—Cockpit production technical specification ready: the choices for cockpit structure and materials should be defined, reviewed and approved.

Module 2, Milestone 2.3—Cockpit production requirements validation: execute dynamic tests for coupling and moving the cockpit with the robot.

Module 2, Milestone 2.4—Cockpit configuration requirements validation: check the simulator components and functions by comparing them with the real aircraft equipment.

Module 3, Milestone 3.1—Visual system technical specification ready: the visual system should be defined, reviewed and approved.

Module 3, Milestone 3.2—Visual system requirements validation: check the resolution, brightness, 3D immersion in compliance with qualifying procedures from FAA Part 60 standard.

Module 4, Milestone 4.1—System Integration specification ready: the integration should be defined, reviewed and approved.

Module 4, Milestone 4.2—Integration requirements validation: quantify the time delay within the whole system and between the subsystems (aerodynamic model response, phased system inputs and outputs, response of the simulator avionics).

Module 5, Milestone 5.1—1st Workshop: kick-off event, when the project objectives were presented and the previous experience from ITA's Manufacturing and Automation Laboratory was demonstrated.

Module 5, Milestone 5.2—2nd Workshop: rotation channel washout filter demonstration using the first SIVOR prototype.

Module 5, Milestone 5.3—3rd Workshop: rotation and translation channels washout filter demonstration using the first SIVOR prototype; presentation of the defined visual system solution architecture.

Module 5, Milestone 5.4—4th Workshop: presentation of the final cockpit configuration mockup.

Module 5, Milestone 5.5—5th Workshop: Cockpit and embedded simulated aircraft systems integration presentation.

Module 5, Milestone 5.6—6th Workshop: Final presentation of the working simulator to the main stakeholders.

Module 6, Milestone 6.1—Robot and track arrival.

Module 6, Milestone 6.2—Visual system equipment arrival.

Module 6, Milestone 6.3—Final cockpit structure arrival.

16.5 VFD's Value Identification Matrix

16.5.1 Stakeholders' Identification + Value Items Analysis

The considered stakeholders were those closely related to the development project. In the case the company (AMC) decides to turn the SIVOR into a commercial product, other stakeholders in the value chain should also be considered. The identified stakeholders and their specific ranking were:

- **Pilots** (primary), which are going to fly the simulator both during the development and after the SIVOR is completed. In this category were included not only the needs from the individual pilots, but also the value pulled by the AMC's flight test group as a whole.
- **AMC** (primary), which represents the knowledge and technology development areas inside of the company and the AMC team that would take part of the project.
- **Regulatory agency** (secondary), which states the FSS certification criteria. If the SIVOR was aiming to develop a commercial product, this stakeholder should be taken as primary; once it is a research project, the team decided to keep it as secondary.
- **Instructor/IOS operator** (secondary), which is the person responsible to configure and operate the simulator during the simulated flights.
- **ITA** (tertiary), which encompasses both the university's development team and the needs from ITA's Manufacturing and Automation Laboratory as a whole, particularly in terms of future use of the acquired knowledge and equipment.

Figures 16.4 and 16.5 present the stakeholder's analysis, including the identified pulled value related to the SIVOR life-cycle. Value identification was accomplished by means of interviews and "go and see", which included visits to flight training centers and EDSs, when the team had the chance to experience the flight in these simulators. Ambiguity from the initially pulled value was eliminated by detailing them into value items (Table 16.2).

Note that the project initial constraints of having an attached to the cockpit visual system was included as a value item, since it is a desirable condition. In the case of using a robotic platform for providing the simulator motion, once it is a must have condition, it does not make sense to treat it as something to be balanced; therefore it was not included as a value item.

As already described in this chapter, FFS types range from Level A to D. On the one hand, stating only being coherent with FAA Part 60 might be ambiguous and confuse the team (a group might work seeking Level D, while other groups might be aiming different targets). On the other hand, arbitrarily defining a particular level to achieve (once no specific level was pulled by the stakeholders) would impose an unnecessary constrains to the project, either setting a too high target or limiting to something below what the team could achieve.

In order to solve this issue a pictorial thermometer-like tool was used, so each team could state its "temperature" in terms of FFS levels (Fig. 16.6). During the

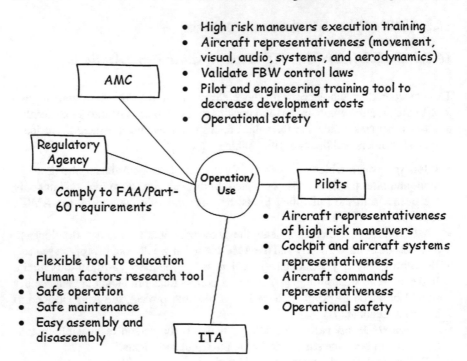

Fig. 16.4 Stakeholders and their pulled values during the product's use/operation

weekly meeting the obstacles to reach higher FFS levels were discussed. Through risk analysis the maximum achievable FFS level was set.

16.5.2 Prioritize the Value Items

From the prioritized stakeholders and the ratings they gave to the pulled value items, Table 16.3 shows the value item's final absolute and relative importance. As a result, the most important value items were related to motion representativeness, aerodynamic model representativeness and safety aspects. During the remaining of the SIVOR development process, the development team kept considering the complete value items' set. The priority was used only when trade-offs were needed.

16.5.3 Define Measures of Effectiveness

After the value items' identification, each value item was associated to an identifiable and measureable effectiveness parameter. The measures of effectiveness allow the identification of the value presence at any moment during the project and are

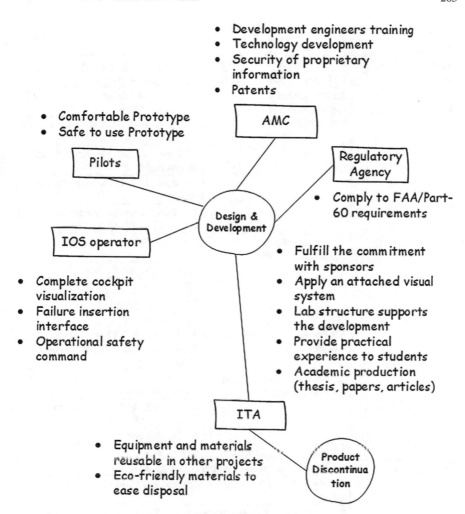

- Development engineers training
- Technology development
- Security of proprietary information
- Patents

- Comfortable Prototype
- Safe to use Prototype

AMC

Pilots

Regulatory Agency

- Comply to FAA/Part-60 requirements

Design & Development

IOS operator

- Complete cockpit visualization
- Failure insertion interface
- Operational safety command

- Fulfill the commitment with sponsors
- Apply an attached visual system
- Lab structure supports the development
- Provide practical experience to students
- Academic production (thesis, papers, articles)

ITA

- Equipment and materials reusable in other projects
- Eco-friendly materials to ease disposal

Product Discontinuation

Fig. 16.5 Stakeholders and their pulled values at design and development, and discontinuation stages

parameters against which the product should be verified during the pull events. Table 16.4 presents some of the chosen measures of effectiveness.

16.5.4 Identify Conflicting Value Items

Some value items may not be optimally delivered simultaneously (Table 16.5). These conflicts are both a challenge and an opportunity to the development team; having a good solution to the conflict might create good opportunities to creating intellectual property assets and patents.

Table 16.2 Filling the VFD—from value to value items

Value	Value item	
1. Execute high risk maneuvers	1.1 Train the pilots to execute on-the-ground high risk maneuvers	
	1.2 Train the pilots to execute in-flight high risk maneuvers	
2. Provide aircraft representativeness	2.1 Provide aerodynamics/movement representativeness	2.1.1 Aerodynamics model coherent with real flight data bank
		2.1.2 Platform motion frequency response coherent with FAA Part 60
		2.1.3 Platform vibration response coherent with real flight
		2.1.4 Ample tridimensional movement workspace
		2.1.5 Platform motion cinematics response coherent with real flight
	2.2 Provide visual representativeness	2.2.1 Attach the visual system to the cockpit
		2.2.2 Cover the pilot's visual filed
		2.2.3 Accurate environment simulation
		2.2.4 High resolution
		2.2.5 3D effect
	2.3 Provide aircraft system' representativeness	2.3.1 Accurate in-flight response
		2.3.2 Accurate cockpit instruments response
	2.4 Provide aircraft cockpit representativeness	2.4.1 Accurate cockpit instruments and flight commands look
		2.4.2 Accurate cockpit instruments and flight commands feel
		2.4.3 Provide comfort
	2.5 Provide aircraft audio representativeness	
3. Flexibility	3.1 Easy/fast to change the aerodynamic model	
	3.2 Possibility of changing the flight commands from conventional to sidestick	
	3.3 Provide a flight instructor workspace to input commands, change simulation parameters and input failures/emergencies	
	3.4 Record simulated flight data	
	3.5 Optimized for the development and test of Fly-by-Wire control laws	
	3.6 Equipment and materials reusable in other projects	
	3.7 Eco-friendly materials to ease disposal	
4. Safety	4.1 Guarantee the safety during the simulation	4.1.1 Guarantee non collision to obstacles in the vicinity of the simulator
		4.1.2 Guarantee pilot's safety inside the cockpit
		4.1.3 Keep simulation acceleration within real normal flight ranges
		4.1.4 Guarantee the simulator safety before starting any simulated flight
		4.1.5 Provide a fast scape alternative

(continued)

Table 16.2 (continued)

Value	Value item	
	4.2 Allow a safe emergency stop of the simulation	4.2.1 Allow the safe simulation stop by the pilot
		4.2.2 Allow the safe simulation stop by the instructor
		4.2.3 Guarantee the safe stop in case of energy failure
		4.2.4 Guarantee the safe stop in case of workspace trespassing
	4.3 Guarantee the safety during maintenance and services	4.3.1 Guarantee the safety against electrical injuries
		4.3.2 Guarantee the safety against mechanical injuries
5. Work as a development tool	5.1 Decrease development costs	
	5.2 Support human factors research	
	5.3 Support engineering training	
	5.4 Support technology development	
	5.5 Generate intellectual property	
6. Security	6.1 Security of proprietary information	
7. Smooth development flow	7.1 Fulfill the commitment with the sponsors	
	7.2 ITA's Manufacturing and Automation Lab supports the development	

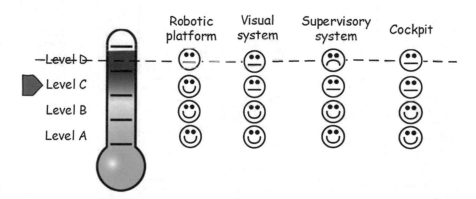

Fig. 16.6 Managing the achievable FFS level

Having the visual system attached to the cockpit increases the challenge to provide both motion and visual representativeness. On the one hand it increases significantly the robot payload, on the other hand it requires the scenario generating algorithms to compensate the joint movement (most of the available algorithms consider a static visual system solution).

Table 16.3 Filling the VFD—value items prioritization

Value Item		Pilots Pri	AMC Pri	Regulatory Agency Sec	Instructor Sec	University Ter	Absolute Importance	Relative Importance
1.1 Train the pilots to execute on-the-ground high risk maneuvers		9	9		3	3	174	3,16%
1.2 Train the pilots to execute in-flight high risk maneuvers		9	9		3	3	174	3,16%
2.1 Provide aerodynamics/ movement representativeness	2.1.1 Aerodynamics model coherent with real flight data bank	9	9	9	9	3	219	3,98%
	2.1.2 Platform motion frequency response coherent with FAA Part 60	3	9	9	1	9	147	2,67%
	2.1.3 Platform vibration response coherent with real flight	3	9	9	1	9	147	2,67%
	2.1.4 Ample tridimensional movement workspace	3	9	3	1	9	129	2,34%
	2.1.5 Platform motion cinematics response coherent with real flight	9	9	9	9	9	225	4,09%
2.2 Provide visual representativeness	2.2.1 Attach the visual system to the cockpit	3	9		1	9	120	2,18%
	2.2.2 Cover the pilot's visual filed	3	3	9	1	9	93	1,69%
	2.2.3 Accurate environment simulation	3	3	9	3	9	99	1,80%
	2.2.4 High resolution	1	1	3		9	36	0,65%
	2.2.5 3D effect	1	1	3		9	36	0,65%
2.3 Provide aircraft system's representativeness	2.3.1 Accurate in-flight response	9	9	9	9	9	225	4,09%
	2.3.2 Accurate cockpit instruments response	9	9	9	9	9	225	4,09%
2.4 Provide aircraft cockpit representativeness	2.4.1 Accurate cockpit instruments and flight commands look	3	3	9		3	84	1,53%
	2.4.2 Accurate cockpit instruments and flight commands feel	9	9	9		9	198	3,60%
	2.4.3 Provide comfort	9	3	9		3	138	2,51%
2.5 Provide aircraft audio representativeness		3	1	9	3	3	75	1,36%
3.1 Easy/fast to change the aerodynamic model			9			9	90	1,63%
3.2 Possibility of changing the flight commands from conventional to sidestick			9			9	90	1,63%
3.3 Provide a flight instructor workspace to input commands, change simulation parameters and input failures/emergencies			3		9	3	57	1,04%
3.4 Record simulated flight data		1	9		9	9	126	2,29%
3.5 Optimized for the development and test of Fly-by-Wire control laws			9			9	90	1,63%
3.6 Equipment and materials reusable in other projects			3			9	36	0,65%
3.7 Eco-friendly materials to ease disposal			1			3	12	0,22%
4.1 Guarantee the safety during the simulation	4.1.1 Guarantee non collision to obstacles in the vicinity of the simulator	9	9	3	3	9	189	3,43%
	4.1.2 Guarantee pilot's safety inside the cockpit	9	9	9		9	198	3,60%
	4.1.3 Keep simulation acceleration within real normal flight ranges	9	9	3		9	180	3,27%
	4.1.4 Guarantee the simulator safety before starting any simulated flight	9	9	3	9	9	207	3,76%
	4.1.5 Provide a fast scape alternative	9	3	9	1	9	147	2,67%
4.2 Allow a safe emergency stop of the simulation	4.2.1 Allow the safe simulation stop by the pilot	9	9			9	171	3,11%
	4.2.2 Allow the safe simulation stop by the instructor	9	9		9	9	198	3,60%
	4.2.3 Guarantee the safe stop in case of energy failure	9	9		9	9	198	3,60%
	4.2.4 Guarantee the safe stop in case of workspace trespassing	9	9		9	9	198	3,60%
4.3 Guarantee the safety during maintenance and services	4.3.1 Guarantee the safety against electrical injuries		9		3	9	99	1,80%
	4.3.2 Guarantee the safety against mechanical injuries		9		3	9	99	1,80%
5.1 Decrease development costs			9			3	84	1,53%
5.2 Support human factors research			9			9	90	1,63%
5.3 Support engineering training			9			1	82	1,49%
5.4 Support technology development			9			3	84	1,53%
5.5 Generate intellectual property			3			1	28	0,51%
6.1 Security of proprietary information			9			3	84	1,53%
7.1 Fulfill the commitment with the sponsors			9			9	90	1,63%
7.2 ITA's Manufacturing and Automation Lab supports the development			3			9	36	0,65%

Table 16.4 Filling the VFD—measures of effectiveness

Value item		Measures of effectiveness
1.1 Train the pilots to execute on-the-ground high risk maneuvers		1.1.1 Perform the following ground test maneuvers: – Ground effect test maneuvers; – Minimum unstick velocity; – Land gear spray; – Ground handling; – Crosswind takeoff; and – Taxiing maneuvers
1.2 Train the pilots to execute in-flight high risk maneuvers		1.2.1 Perform the following in-flight test maneuvers: – Auto-off; – Stall performance; – In-flight refueling; – Engine failure after takeoff; and – Flutter test
2.1 Provide aerodynamically/ movement representativeness	2.1.1 Aerodynamic model coherent with real flight data bank	2.1.1.1 Average error <x % 2.1.1.2 Environment simulation (atmosphere, wind and turbulence)
	2.1.2 Platform motion frequency response coherent with FAA Part 60	2.1.2.1 Bandwidth > x
	2.1.3 Platform vibration response coherent with real flight	2.1.3.1 Generated vibration > 10 Hz
	2.1.4 Ample tridimensional movement workspace	2.1.4.1 Workspace envelope $\geq$ 79.8 m^3
	2.1.5 Platform motion cinematics response coherent with real flight	2.1.5.1 Positive feedback from experienced pilots

Table 16.5 SIVOR's conflicting values considered in the YET analysis

Primary value		Secondary value
2.1 Provide aerodynamics/movement representativeness	YET	2.2.1 Attach visual system to the cockpit
2.4 Provide aircraft cockpit representativeness	YET	3.2 Possibility of changing the flight commands from conventional to sidestick

In the same way, providing both aircraft commands' representativeness (as in FAA/Part-60 requirements) as well as interchangeability is not trivial.

These two aspects challenge directly the project's main goal of having both the features from a FFS and an EDS, also capable of simulating high risk maneuvers.

16.6 Study Phase—Value Proposition Activities

16.6.1 New Product Design: Functional Architecture

Considering the identified value set, the team designed a functional architecture capable of delivering it. Functional analysis was applied (see Chap. 11) resulting in the functions presented in Fig. 16.7. Note that it included functions related not only to the SIVOR operation, but also to maintenance and to the needed ITA Lab's support and infrastructure in order to guarantee the housing, safety and security of the project equipment and information.

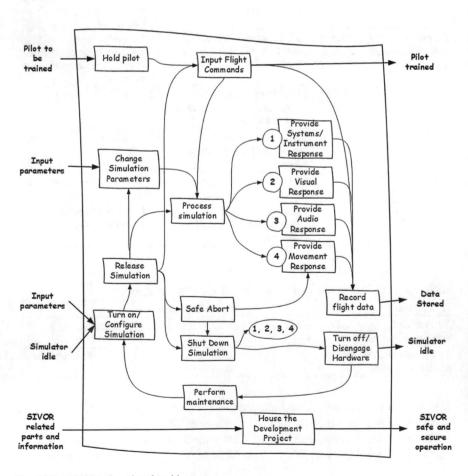

Fig. 16.7   SIVOR's functional architecture

16.6.2 Risks and Issues

The team identified and analyzed the likelihood and impact (Fig. 16.8) of seven risks they decided to address during the development execution. These risks might prevent the value items delivery by the designed functional architecture:

Risk 1. Embedded visual system exceeds 250 kg;
Risk 2. Mechanical structure does not achieve the needed rigidity (vibration mode above 10 Hz);
Risk 3. Flight commands do not give representative force feedback;
Risk 4. Main components of the cockpit (throttles, column, instruments) not available;
Risk 5. Emergency stop system does not fulfil minimum safety regulations;
Risk 6. The position of the cockpit structure's CG (center of gravity) get too far from the robot flange; and
Risk 7. Incomplete integration between the software simulation platform and the aircraft's available flight model.

16.6.3 Fill the VFD's Value Rework Avoidance Sub-matrix

The functions identified during the functional analysis were transported to the VFD and related to the value items they support delivering. On Table 16.6, the value from column "Absolute Importance" (as in Table 16.3) is proportionally distributed among the related value delivery functions. Taking, for instance, the value item "1.1 Train the pilots to execute on-the-ground high risk maneuvers," the absolute importance (174) was divided by the sum of the weight from each related function $(3 + 0 + 3 + 1 + 3 + 9 + 9 + 3 + 3 + 9 + 0 + 0 + 0 + 0 + 0 = 43)$, which resulted in shares of 4.05. Finally, the absolute importance is distributed to each function according to its supporting contribution.

Risk	Likelihood (1-5)	Impact (1-5)	Total	Grade
1	2	4	8	Medium
2	2	4	8	Medium
3	2	3	6	Low
4	1	4	4	Low
5	1	5	5	Low
6	2	4	8	Medium
7	3	4	12	Medium

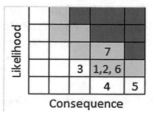

Fig. 16.8 Identified risks' likelihood and impact

Table 16.6 Relating value and risk to each value delivery function

Value	Value Item	Absolute Importance	Turn on/ Configure Simulation	Release Simulation	Change Simulation Parameters	Hold Pilot	Input Flight Commands	Process Simulation	Provide Systems/ Instrument Response	Provide Visual Response	Provide Audio Response	Provide Movement Response	Safe Abort	Shut Down Simulation	Record Flight Data	Turn off/Disengage Hardware	Perform maintenance	House the Development Project
1.Train the pilots to execute high risk maneuvers / 1. Train the pilots to execute high risk maneuvers	1.1 Train the pilots to execute on-the-ground high risk maneuvers	174	3		3	1	3	9	9	3	3	9						
			12,1		12,1	4,0	12,1	36,4	36,4	12,1	12,1	36,4						
	1.2 Train the pilots to execute in-flight high risk maneuvers	174	3		3	1	3	9	9	3	3	9						
			12,1		12,1	4,0	12,1	36,4	36,4	12,1	12,1	36,4						
...	...	...	...	...	...	...	...	...	...	...	...	...	...	...	...	...	...	...
R1 - Visual embedded system exceed 250kg		8	0	0	0	0	0	0	0	8	0	0	0	0	0	0	0	0
R2 - Mechanical structure does not achieve the needed rigidity		8	0	0	0	0	0	0,8	0	0	0	7,2	0	0	0	0	0	0
...		...	...	...	...	...	...	...	...	...	...	...	...	...	...	...	...	...

Besides the value items, the value delivery functions were also related to the identified development risks. From this linking, the team considered the results from the likelihood X impact analysis and used the same distributive procedure as the one for distributing shares from the value items' absolute importance (Table 16.6).

Figure 16.9 shows the best candidate functions to SBCE. These functions are the ones that combine a greater total value (TV) and perceived risk (PR) to successfully deliver this value. These are the functions which the development team should seriously consider to carry out multiple alternatives during the development project.

There are two critical functions:

(1) "Provide Movement Response", since using a robotic motion platform to accurately simulate high risk maneuvers, particularly by having the added track, is the core project innovation; and
(2) "Process Simulation", once it relates to several subsystems: the simulation engine, the aircraft flight model, all the interfaces with instruments, the motion platform, the audio system, and the video system.

16.6.4 Finding Alternatives

Even though the most critical value delivery function was the "Provide Movement Response", it does not make sense to pick it to explore alternatives, once not having the robot and the track is not an option. As a consequence, the team decided to

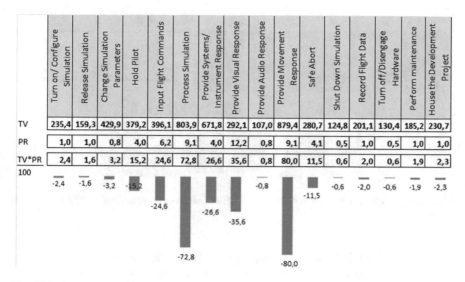

Fig. 16.9 Best candidate functions to SBCE

consider the assembly of the robot and the visual system. Besides being the third function in terms of criticality, the visual system has the greatest potential to push the motion to the borders of its working envelope, therefore challenging its performance and value delivery capacity.

The SBCE alternatives for the set ("Provide Movement Response" + "Provide Visual Response") were defined through brainstorming sessions. Eight different ways of performing the visual response were figured out (Fig. 16.10) from that:

A1.   Embedded FFS Level D capable system: this is the team's preferred choice.
A2.   Embedded FFS Level C capable system: the required 3D capabilities are reduced; consequently, can use less weighting projectors.
A3.   Embedded FFS Level B capable system: very simple and light visual system. Just one projector with field of vision of 45° horizontal and 30° vertical.
A4.   Embedded FFS Level B capable system with back projection: the screens are put closer or attached to the windows and the projection comes from the back. The proximity between the pilot and the screen reduce the depth of field, therefore limiting this alternative to FFS Level B.
A5.   Embedded FFS Level D capable system with back projection and mirrors: the screens are put closer or attached to the windows and the projection comes from the back. By using mirrors the projectors can be installed closer to the cockpit, reducing the mechanical stress of the robotic platform, keeping the CG closer to the robot flange, and requiring less workspace to the SIVOR as a whole. The mirrors, though, increase cost and weight.

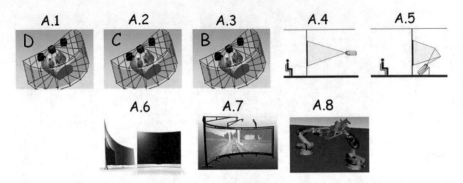

Fig. 16.10 SBCE alternatives for the set "Provide Movement Response" + "Provide Visual Response"

A6. Embedded FFS Level B capable system with OLED (Organic Light-Emitting Diode) screens on the windows: use of curved OLED screens close to the cockpit windows. Brings all the benefits from alternative 5 and weighs less than the others.

A7. External FFS Level D capable system: this is the traditional solution, where the projection system and the cockpit + motion are completely separate; it restrains the workspace and requires a dedicate environment/building to the simulator, reducing its flexibility.

A8. External FFS Level D capable system using cooperative robotic platforms: this alternative considers the use of two cooperative robots, one holding the cockpit and the other the visual system; during operation they should work coordinated as a "team". This alternative imposes higher complexity and cost than all of the previous ones.

In order to determine the alternatives that best fit the value set, the team made a Pugh analysis (positive [1], negative [−1], or neutral [0] correlation) verifying how each alternative behaved against the pulled value. Table 16.7 shows the results of the analysis. Alternatives 7 and 8 were not considered for analysis: the former does not fit in the current Lab size and would require a new building; the latter dangerously increases the cost (beyond the budget) and complexity of the project. It can be, though, the scope of a follow up research project.

Among all the solutions, the use of an A1.Embedded FFS Level D capable system got the highest score, although is the one that imposes higher risks. Analyzing the value delivery capacity versus related risks to each alternative (Fig. 16.11), the team set the strategy of keeping alternative 1, 2, 3 and 6. Once alternative 1 can be downgraded to 2 and 3 with minimum needed rework and new equipment acquisition, keeping all the three in mind gives the team flexibility to not fail in delivering, within the project timeframe, a video system, even if it is not the most capable one. Alternative 6 is going to be developed in parallel, until the risks R1, R2 and R6 are mitigated.

Table 16.7 "Provide Movement Response" + "Provide Visual Response" alternatives analysis

Value Items	A1	A2	A3	A4	A5	A6
1.1 Train the pilots to execute on-the-ground high risk maneuvers	0	0	0	0	0	0
1.2 Train the pilots to execute in-flight high risk maneuvers	0	0	0	0	0	0
2.2.1 Cover the pilot's visual filed	1	0	-1	1	1	1
2.2.2 Accurate environment simulation	1	0	-1	-1	0	1
2.2.3 High resolution	1	0	0	0	0	1
2.2.4 3D effect	1	0	-1	-1	0	0
3.6 Equipment and materials reusable in other projects	0	0	0	0	-1	1
3.7 Eco-friendly materials to ease disposal	0	0	0	0	0	0
4.1.2 Guarantee pilot's safety inside the cockpit	1	1	1	0	0	-1
4.3.1 Guarantee the safety against electrical injuries	0	0	0	0	0	0
4.3.2 Guarantee the safety against mechanical injuries	0	0	0	0	0	0
5.1 Decrease development costs	0	0	0	0	0	0
5.2 Support human factors research	1	0	0	0	0	0
5.4 Support technology development	0	0	0	0	0	0
5.5 Generate intellectual property	0	0	0	0	0	0
	6	1	-2	-1	0	3

	A1	A2	A3	A4	A5	A6
R1 - Visual embedded system exceed 250kg	-1	0	0	0	0	1
R2 - Mechanical structure does not achieve the needed rigidity (vibration mode above 10Hz)	-1	0	1	1	1	1
R6 - Structure's CG position get too far from the robot flange	-1	0	1	-1	0	1
R7 - Incomplete integration of the software simulation platform and the aircraft flight model	0	0	0	0	0	0
	-3	0	2	0	1	3

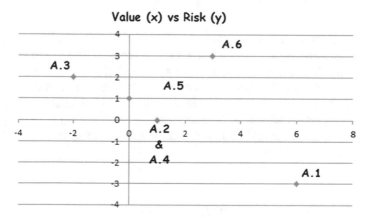

Fig. 16.11 Value versus risk graph

Table 16.8 Process simulation alternatives

	Simulator engine	Integration environment	Aircraft model
A1	Simulink©	AMC's IronBird ©	AMC's aircraft model
A2	Simulink©	LabVIEW©	AMC's aircraft model
A3	XPlane©	XPlane©	XPlane's compatible model
A4	Other flight simulator but Xplance©	Other flight simulator but Xplance©	A compatible and free to use aircraft model
A5	LabVIEW©	LabVIEW©	Use Matlab© to develop a model from scratch

In the case of the "Process Simulation" value delivery function, the considered alternatives were defined by varying possible simulating engines, aircraft models, and environments to integrate the simulation engine and the aircraft model (Table 16.8), where:

A1.   Use AMC's virtual aircraft architecture: this is the team's preferred alternative, once it gives higher chance to achieve the requirements from FFS level D.

A2.   Use the software LabVIEW® to integrate AMC's aircraft model. In this case the Level D might be loss due to integration limitations.

A3.   Use the commercial flight simulator software XPlane®, Simulink®, and an aircraft model that best resembles the real aircraft model. This alternative limits the solution to achieving FFS level B.

A4.   This alternative is similar to A3, but uses any other flight simulator package but XPlane®. Besides carrying the same drawbacks from the A3, the team would need to learn the particularities from another software package.

A5.   The team would build its own aircraft model to integrate using LabVIEW® and Matlab®.

The team decided to keep alternatives A1, A2 and A3. They already implemented and used alternative A3 during the initial tests; it though limits the SIVOR to FFS level B. Alternatives A1 and A2 might open the possibility to FFS level D.

Alternatives A4 and A5 were dropped, once they require a great effort from the development team, either by having to learn another flight simulator package (they already have good domain of the XPlane® engine), or by having to develop an aircraft model from scratch, and with minimum chance of performing better than the XPlane's already available ones.

16.7 Study Phase—Planning Activities

In order to better give priority and to avoid redundant activities, the responsibilities related to each value delivery function were defined and represented in an RACI chart at the top of the VFD (Table 16.9).

Table 16.9 SIVOR RACI chart

	Turn on/Configure Simulation	Release Simulation	Change Simulation Parameters	Hold Pilot	Input Flight Commands	Process Simulation	Provide Systems/Instrument	Provide Visual Response	Provide Audio Response	Provide Movement Response	Safe Abort	Shut Down Simulation	Record Flight Data	Turn off/Disengage Hardware	Perform maintenance	House the Development
Management	A	A	A	A	A	A	A	A	A	A	A	A	A	A	A	R/A
Maintenance and support			C	I			I	I	I	I	I	I		I	R	C
Supervisory system	R	R	R		R	R	R	I	C	C	C	R	R	R	C	C
Visual system			C		I		R		C			C		C	C	C
Cockpit				R	C		C	C	R	C	C		C	C	C	C
Robotic platform		R		C		C	I	I			R	R	C	C	R	C

16.7.1 Fill the VFD's Flow Definition Sub-matrix

All but milestone 5.1 (which was the project kick-off meting) from the initially defined project milestones were kept as pull events and considered the objectives as presented in Table 16.10, where:

- Analysis (A): Qualitative analytical evidence is given that the actual state of the solution embeds the related pulled value.
- Calculus (C): Quantitative evidence is given that the actual state of the solution embeds the related pulled value.
- Inspection (I): Inspection of the actual state of the solution shows that it embeds the related pulled value.
- Demonstration (D): Demonstration of the actual state of the solution shows that it is capable of delivering the related pulled value.
- Test (T): By testing the actual state of the solution it is shown that the related pulled value is present.

The pull events were then detailed into activities, which were put in the progress board's backlog for weekly progress tracking. It's important to note that the workshops acted themselves as pull events, due to the fact that the project's sponsors and other key stakeholders were always invited.

16.8 Execution Phase—Design and Development Activities

Some of the IPD tools have great potential impact during the execution activities. First of all, it is necessary to consider two crucial integrative design variables:

(1) weight, which relates to the maximum robot payload; and
(2) center of gravity (CG) location, which is also limited by the robot characteristics. These two variables drive the use of two IPD techniques: Design to Weight (DTW) and Design to CG (DTCG). The pulled value of flexibility, particularly in terms of exchanging flight commands and instrument, brings the need of applying the Design for Assembly (DFA) directives. Finally, the safety concerns pull the use of Design for Safety (DFS).

Table 16.10　SIVOR pull events' scope

Value Item (check description in table 16.2)		1.1	1.2	2.1	2.2	2.3	2.4	3.1	3.2	4.1	4.2	5.2	5.3	5.4	5.5	5.6	6.1	6.2
1.1		A	A	A			A	A	A	A	D	D	T	A	D	T		
1.2		A	A	A			A	A	A	A	D	D	T	A	D	T		
2.1	2.1.1									A	D	D	D	D	D	T		
	2.1.2	A	A									D	T	D	T	T		
	2.1.3	A	A									D	D	D	T	T		
	2.1.4	A	C									D	T	D	T	T		
	2.1.5											D	T	D	T	T		
2.2	2.2.1			A	A	C	C	A	C					D	T	T		
	2.2.2			A			A	A	D					D	T	T		
	2.2.3							A	D					D	T	T		
	2.2.4							A	D					D	T	T		
	2.2.5							A	D					D	T	T		
2.3	2.3.1									A	A	D	D	D	T	T		
	2.3.2			A			A			A	A	D	D	D	T	T		
2.4	2.4.1			A			I							T	T	T		
	2.4.2			A			I					D	D	D	T	T		
	2.4.3			A	A	I		A	T					D	T	T		
2.5				A	A	I	I			A	A	D	D	A	T	T		
3.1										A	D	D	D	D	D	T		
3.2				A	A	D	D								D	T		
3.3										A	D	D	T	T	T	T		
3.4		A	A	A			A			A	D	D	T	T	T	T		
3.5		A	A							A	D		A	D	T	T		
3.6		A	A		A	D		A	D	A	A				A			
3.7								A	D						A			
4.1	4.1.1							A	A			T	T	T	T	T		
	4.1.2			A	A	C	C	A	A	A	A	D	D	D	T	T		
	4.1.3									A	A	D	D	T	T	T		
	4.1.4									A	D	D	D	T	T	T		
	4.1.5			A	A	I	I	A	A				A	D	T	T		
4.2	4.2.1	A	A							A	D	D	D	D	T	T		
	4.2.2	A	A							A	D	D	D	D	T	T		
	4.2.3	A	A							A	A	D	D	D	T	T		
	4.2.4	A	A							A	A	D	D	D	T	T		
4.3	4.3.1	A	A					A	A			A	A	D	D			
	4.3.2	A	A	A	A	I	I	A	A			A	A	D	D			
5.1										A	A	A	A	A	D	D		
5.2							A			A	A	A	A	A	D	D		
5.3										A	A	A	A	A	D	D		
5.4		A	D	A	A	D	D	A	D	A	D	D	D	D	D	D		
5.5		A	D	A	A	D	D	A	D	A	D	D	D	D	D	D		
6.1										A	D	T	T	T	T	T		
7.1		A	A	A			I	A	I	A	C	D	D	D	T	T		
7.2		A	A	A			I	A	I	A	I	T	T	T	T	T	T	T
Risks (check description in section 16.6.2)																		
R1								A	C					D	T	T		
R2		A	A										D	T	T	T		
R3				A			D								T	T		
R4														D	T	T		
R5													D	T	T	T		
R6				A	A	C	C	A	C				C	D	T	T		
R7										A	A	A	A	D	T	T		

16.9 A Practical View

Like the SIVOR, we often have our development constrained by early project decision. When answering a bid, a series of product decision restrictions are given. In contracted developments, it is very common that we sign it with clauses that embed a lot of wishful thinking. The world is what it is.

This chapter gives a good example of how creatively dealing with project constraints and yet adapting and using the method presented in Chaps. 9–13:

1. Adaptation of an initial wishful thinking schedule into milestones and pull events without losing track from what has already been celebrated with the project sponsors.
2. Managing and guaranteeing the unambiguity from a flexible value set.
3. Finding value delivery functions possible alternatives that obey the initial constraints, delivery value, and reduce risks.

In the case the team decides to evolve the prototype into a commercial product, some stakeholders should be promoted (the regulatory agency should go from secondary to primary), and other stakeholders in the value chain should be considered (such as those related to installation and maintenance services).

The functional architecture should include the installation service and detail further the maintenance services. Imagine that the additional pulled value drives the company to offer maintenance services guaranteeing the client the product availability; In this case functions such as "monitor the product performance", "plan predictive/preventive maintenance", and "perform the maintenance" might be added.

Appendix A
Product Development Performance Drivers[1]

This appendix details the product development performance drivers groups, as presented in Chap. 1, into categories and subcategories.

A.1 External Environment

The external environment is divided into two categories: market and business.

A.1.1 Market

Market includes aspects related to the maturity of the product's design on the market, consumer decision, globalization and product lifecycles. This category has four subcategories:

Market: No dominant design of product. The lack of a dominant design means that the market has not yet made up its mind about the product. A dominant design is the one that wins the allegiance of the marketplace and those competitors and innovators must adhere, making many of the performance requirements implicit in the design itself. If there is still no dominant design, intensive churning of product innovations is expected.

Market: Inability to understand customer decision. Customer change-decision logic is beyond the company's comprehension, leading to constant requirement's change, work in process, information obsolescence, and its rework consequences.

---

[1]Adapted from: Pessôa, MVP (2008) Weaving the waste net: a model to the product development system low performance drivers and its causes. Lean Aerospace Initiative Report WP08-01, MIT: Cambridge, MA.

© Springer International Publishing AG 2017
M.V.P. Pessôa and L.G. Trabasso, *The Lean Product Design and Development Journey*, DOI 10.1007/978-3-319-46792-4

Market: Global markets. Global markets impose different sets of regional requirements which may be difficult to manage (volume), to define trade-offs (solving conflicts), and to maintain consistency (data/unit system conversions).

Market: Decreasing product life cycles. Decreasing product life cycles is a reality in most (if not all) industries. Sometimes the life cycles even become (much) smaller than the PD lead time. Even though a dominant design exists, innovations from competitors (frequently) shake the market, turning PD work in process obsolete.

A.1.2 Business

The business category includes instabilities in the business scenarios, as shown in the following subcategories:

Business: Manifold laws and restrictions. The business environment is constrained by several laws and restrictions which constitute high risk to the industry since: (1) they have a significant impact; (2) their change probability is high; or (3) both.

Business: Patents. Competitors use patents to create "mine fields" and limit the company's alternatives to development.

Business: Changes on the political scenario. Instability and changes on the political scenario constitute high risk to the industry since: (1) they have a significant impact; (2) their change probability is high; or (3) both.

Business: Changes on the economic scenario. The industry's products elasticity is (very) high, thus, instability and changes on the economic scenario have great impact on the company.

Business: Changes on the social scenario. Instability and changes on the social scenario historically constitute high risk to the industry since: (1) they have a significant impact; (2) their change probability is high; or (3) both.

Business: Labor factors. Labor factors historically constitute high risk to the industry, or are expected to impact the company in the near future.

Business: Ecological/Environmental factors. Ecological/environmental factors historically constitute high risk.

A.2 Internal Environment

The internal environment includes everything that is outside of the PDS but is still within the boundaries of the parent organization. The internal environment was divided into five categories: organizational culture, corporate strategy, organizational structure, business functions, and supporting processes.

A.2.1 Organizational Culture

Organizational culture is the culture that exists in an organization. It is made up of such things as values, beliefs, assumptions, perceptions, behavioral norms, artifacts, and patterns of behavior. It is a socially constructed, unseen, and unobservable force behind organizational activities. It is a social energy that moves organization members to act. It is a unifying theme that provides meaning, direction, and mobilization for organization members. It functions as an organizational control mechanism, informally approving or prohibiting behaviors [1]. Therefore, the organizational culture plays a critical and role on how the PDS is really structured and executed, sometimes in different ways than the company's standards.

Culture: Failure is likely to be blamed on individuals. Corporate culture leads to witch hunt on every failure instead of taking the learning opportunity.

Culture: Reluctance to accept bad news. "Kill the messenger." Bad news is either ignored or not seriously considered ("we know better"). The person who raises the issue is taken as the problem.

Culture: Bad news "softening". Bad news is softened as it is passed to higher levels on the hierarchy structure.

Culture: Lack of strong enforcement of the schedule. There is no culture to relentlessly enforce the committed schedule. Delays impact the project and may disturb the whole portfolio resource management and break all the commitments.

Culture: Lack of appropriate incentive (meritocracy). The incentive system criteria are detached from final customer value delivery metrics.

Culture: No "listen to"/consent=weakness. Listening and considering subordinate's opinions is considered management weakness.

Culture: No open-minded teamwork culture. Team members are not motivated/ allowed to purpose improvements or to present their opinion.

Culture: No "learning DNA," no learning culture. Company's culture does not foster the knowledge capture, dissemination, and reuse. Even if there is a knowledge management system in place, there is no "learning DNA" to make use of it.

Culture: Micro-policies and hidden personal targets. It is possible that personal targets overcome corporate objectives, even if they are conflicting.

Culture: Competitive climate. Excessive focus on the individual success, rather than on the group acknowledgements. A too competitive climate may prevent knowledge sharing and the creation of high performance teams.

A.2.2 Corporate Strategy

According to Andrews [2], Corporate Strategy is the pattern of decisions in a company that determines and reveals its objectives, purposes, or goals, produces the principal policies and plans for achieving those goals, and defines the range of business the company is to pursue, the kind of economic and human organization it is or intends to be, and the nature of the economic and non-economic contribution it intends to make to its shareholders, employees, customers, and communities.

Unclear strategies or the misalignment between the corporate strategy and the development needs and goals is a factor that reduces development performance.

Strategy: Lack of solid strategy. Corporate strategy changes frequently and "randomly." The strategy might even be very well defined and communicated, but the frequent changes make people uncertain about the future and skeptic to changes.

Strategy: Missing or rather unclear objectives/targets. Strategy does not define or unclearly defines its objectives/targets leading to disconnected efforts and/or ambiguous interpretations.

Strategy: No forward-looking knowledge of market potential. Limited search to new markets and new uses for products and technologies. The lack of market potential knowledge may lead to the design of low flexibility products, consequently reducing the expectation of future reuse.

Strategy: Bad outsourcing. Outsourcing involves the transfer of the management and/or day-to-day execution of an entire business function to an external service provider. Non-core operations are the focus of outsourcing. Considering the business functions that interface with the development, bad outsourcing impacts negatively on the product development process execution.

Strategy: No strategic product architecture. Product architectures are defined to the short term, leading to poor future reuse.

Strategy: Technology development concurrent with development of product. Company assumes the risk of developing both technology and product at the same time the product is being developed. The lack of an already capable technology may lead to the definition of patch solutions.

Strategy: No forecasting roadmap in technology. The company does not have a technology roadmap in order to be ahead of competition.

Strategy: No forecasting roadmap in manufacturing. The company does not have a manufacturing roadmap in order to support the evolution of the technologies, causing manufacturing to be always behind and reactive. As a side effect the company does not take advantage of possible manufacturing evolutions and greater efficiency.

Strategy: Low commitment to work environment (employee well-being). Employee well-being is secondary. Symptoms are low motivation, low commitment, and high turnover.

Strategy: Confidentiality of expertise. Expertise is not shared due to confidentiality. The higher the amount of expertise considered confidential, the more compromised are the knowledge sharing and the prevention of re-invention.

Strategy: Too many projects (over-utilization). Project teams are over-utilized due to multitasking between projects or/and functional tasks.

A.2.3 Organizational Structure

The organizational structure defines responsibilities, authorities, and relations in order to enable the performing of organization functions including the product development.

Structure: Project leader has no real power. The company gives limited power to the project manager (high accountability but low power to influence the company).

Structure: Scattering structure. A scattering structure disrupts the subtle interactions required for teamwork, turning the knowledge flow ineffective [3]. Scattering organizational structures are: (1) confusing due to unclear interfaces; and/or (2) complex as the result of multiple/intricate interfaces. Excessive decentralized structures may be an example.

Structure: Unclear or mismatching policies, roles, and rules. Unclear or mismatching policies, roles, functions, and rules enforce scattering by disrupting and turning ineffective the knowledge flow) [3]. The "how to behave" is not clear, thus, leading to unpredictable actions.

Structure: Excessive conservatism and bureaucracy. Company has a stovepipe structure, implying long decision and lead times.

Structure: Excessive specialization of tasks. The specialization of tasks makes it difficult for the team members to have a system view. People do not figure out how their job contributes to the whole, thus leading to local optimizations rather them the system optimization.

Structure: Mismatch between responsibility and rights. The mismatch between responsibility and rights prevents motivation and commitment.

Structure: High interdependencies. Structure is strongly coupled and inflexible.

Structure: Poor empowerment. Centralization of decision creates low flexibility and long lead-times.

Structure: Low technical knowledge of executives. Executive's low technical knowledge makes it difficult for them to understand the product development issues.

A.2.4 Business Functions

The business functions are: human resources, sales and marketing, research and development, production/operations, customer service, finance and accounts, and administration and information technology. This category considers the issues between the product development and the other business functions in the company.

Functions: Business functions are not integrated to PD. Business functions commitment is not part of the PD process, thus, their knowledge and restrictions are not considered during the design. Some examples are: (1) Manufacturing does not participate during PD; (2) PD does not design future service and support; (3) Sales presence is largely absent during PD cycle; etc.

Functions: Functions have long or unpredictable lead times. Business functions have long or unpredictable lead times, delaying the development or creating huge buffers in schedules.

Functions: No/low commitment from business functions to PD. Even though invited to participate in the development process, the business functions do not actively contribute. Lack of commitment may result in later changes in the development goals and agreements, delaying the end of the product development phase.

Functions: Functions have quality problems. Internal functions deliverables present quality problems.

Functions: Warranty issues. Suppliers' deliverables present quality problems.

Functions: Very formal ties between PD and supplier. PD team has bounded access to suppliers not allowing a "four handed" development. Suppliers do only what has been contracted and no other opportunities are explored during the development process. There is no partnership from suppliers.

A.2.5 Supporting Processes

The supporting processes support another process(es) as an integral part with a distinct purpose and contribute to the success of the development project. The supporting processes considered in this category are: improvement, training and knowledge management.

Proc Improvement: improvement detached from current process understanding and measures. Process improvement not based on a thorough understanding of the current strengths and weaknesses of the organization's processes and process assets.

Proc Improvement: no systematic continuous improvement. Measurable improvements to the organization's processes and technologies are NOT continually and systematically deployed.

Measurement: no or impractical measurement system. Project metrics do not support corrective and proactive improvement actions: (1) measurement objectives are not aligned with the company's objectives and related information needs; (2) measures do not address the measurement objectives; (3) procedures for measurement data analysis and reporting are not defined; or (4) relevant stakeholders do not receive the results of measurement and analysis activities.

Training: not defined training curriculum. The lack of a defined training curriculum prevents the creation of a leveled knowledge base on the resource pools, impairing the communication and the sharing of experiences.

Training: delivered training does not match with planned. Organizational training process cannot guarantee that the delivered training match the needs.

Offered training is more "waste of time" than really an opportunity to give useful knowledge to the PD team.

Training: training detached from organizational needs. Given training do not develop the skills and knowledge of people so they can perform their roles effectively and efficiently. Training not attached to the organization's strategic needs.

Knowledge: no/poor knowledge management system. No/ poor system for data collection, management and reuse.

Knowledge: no archiving discipline. No/poor definitions of who, what, when and where store information resulting on: (1) storing of useless data; (2) multiple sources for the same information; (3) private databases; and (4) collection and storage of "just-incase" information.

Knowledge: poor Know-How and tacit knowledge transfer. No poor tacit knowledge transfer in order to avoid repeating mistakes, and to allow repeating successes.

Knowledge: low re-use rate of physical and design assets. People are not enforced to reuse physical and design assets from previous and successful developments.

Knowledge: low reliability of information. Information on knowledge base is not reliable (wrong, incomplete or obsolete) leading to lack of confidence and rework.

Knowledge: hard to find information. Heritage information is difficult to find/ access, preventing its future use.

Knowledge: critical information not available. Data, answers, decisions (review events) are not available on knowledge base.

A.3 Project Environment

This source encompasses all the product development management and execution activities and is divided into six categories: initiation, development planning, execution management, development control, communication, and development execution [4].

A.3.1 Initiation

The initiation defines and authorizes the development. The initiation must guarantee the alignment between the development and the corporate strategy through clear and feasible objectives.

Initiation: Project objectives are narrowly defined and/or unclear. Project goals are narrowly defined (does not exploit the whole set of potential benefits) and/or unclear (the core benefits to be achieved are fuzzy).

Initiation: Project misaligned from corporate objectives or corporate values. There is some sort of misalignment between the project goals/means and the company's strategy, objectives, or values.

Initiation: Product positioning is not customer related. Product positioning strategy is rather driven by internal assumptions and issues (brands, trade, higher management assumptions, etc.) than based on the true customer needs.

Initiation: Bad product costing and pricing strategy. Costing and pricing strategies are not driven by the final customer paying willingness.

A.3.2 Development Planning

The planning defines and refines objectives, and plans the course of action required to attain the objectives and scope that the project was undertaken to address. The development planning subcategories are:

Planning: Standard processes not followed. The defined process for the project does not follow the company's standard process.

Integration: Inconsistence between plans or plans' parts. The inner parts of the plan have conflicting information leading to confusion or multiple interpretations.

Scope: Unclear/partially defined. Project work and expected deliverables are not clear (do not know what to do).

Time: Schedule is too long. Excessive activities' duration. Excessive buffers.

Time: Schedule is too compressed. Exiguous time to execute the activities or excessive fast tracking (parallel activities), leading to a high risk level.

Cost: Bad budget definition. Budget estimated without criteria (a bet).

Quality: Bad quality planning. Quality planning (the identification of which quality standards are relevant to the project and determining how to satisfy them) not done or poorly done.

Human Resource: Poor resource allocation. Resources are allocated not considering the activities prerequisites and the actual resource knowledge/experience.

Human Resource: No resource leveling. Unbalanced uses of resources (usually people). Over time, over-allocations, and conflicts are not resolved.

Human Resource: No/poor stakeholder commitment and involvement. Relevant stakeholders are not involved in the planning process. Critical dependencies are not identified, negotiated, and tracked. Commitment from relevant stakeholders responsible for performing and supporting plan execution is not obtained.

Communications: Poor communications planning. Communication planning (who, how, what, and when will receive communications) not or poorly done.

Risk: No/poor risk identification and response planning. Not doing or doing bad risk identification and response planning. Taking risks beyond the necessary.

Procurement: Poor procurement planning. Procurement planning not or poorly done, including purchase planning, and contract's type definition.

A.3.3 Execution management

The execution management integrates people and other resources to carry out the planned project for the project. The execution management subcategories are:

Execution: Missing or not followed plan. There is no plan or the existent plan (good or bad) is not followed (bad planning is related to the "project planning" group above).

Execution: Doing without knowing or understanding. People do not know or do not have the right understanding of what is expected from their individual tasks.

Execution: Priorities not clearly defined. Whenever there is a lack of resources/time there is not a clear priority to allocate resources or define what to do first.

Execution: Resource availability below the demand. Available resources are insufficient to fulfill the demand.

Execution: System over utilization. The system stresses its resources due to over utilization.

Execution: Multitasking. Resource executes multiple tasks on the same project and/or outside the project (on other projects or doing functional activities).

Execution: Poor knowledge transfer. Knowledge acquired during execution is not spread efficiently to expedite execution and to promote mistake avoidance.

Execution: Inadequate information delivered. Not ready or wrong information is delivered during execution. This includes both information deliverables and decision data.

Execution: Considered only "inside the fence" alternatives. Development team does not consider outside sources of products that may be used to satisfy the project's requirements. Suppliers' capability is not exploited. Of-the-shelf solutions are not seriously considered.

Execution: Undisciplined processes/work. The work done does not follow the standard procedures/guidelines.

Execution: Bad information handoffs. Information is lost during information transfer due to bad handoffs.

Execution: Lack of shared vision. The team members do not have a holistic view and are not capable to see how their individual work contributes to the whole.

Execution: No/ineffective corrective actions. Corrective actions to identified problems are not executed or are inefficient (they do not address the root causes).

Execution: Poor receiving and transition products from suppliers. Products acquired from suppliers are not conformance checked at arrival, and/or are not timely distributed to the development team.

Execution: Inefficient supplier selection systems. Supplier selection system prevents the right trade-off between price and quality: (1) lower price wins regardless of quality; (2) buy from preferred suppliers "regardless" of price; (3) buy from "friends."

Execution: Product service processes is a low level priority. The development of the product service processes is overlooked during development. The service requirements are not considered at the beginning of the development.

Execution: Bad cash flow. Actual cash flow does not meet the budget needs and timing of the development projects.

Execution: Poor change management. Changes are badly managed, controlled, and informed: (1) changes are hardly traceable to who requested and authorized; or (2) people impacted by the change are not timely informed.

Execution: Poor WIP version management. No/poor WIP version management system causing confusion on which version is the right to use: (1) there are no defined baselines for internal use and for delivery to the customer; (2) the configuration items, components, and related work products that will be placed under configuration management are not identified or badly described; (3) there is no configuration management and change management system for controlling work products; or (4) the integrity of the configuration baselines is not maintained.

A.4 Development Control

The development control regularly measures and monitors progress to identify variances from the project management plan so that corrective actions can be taken when necessary to meet project objectives. The development control subcategories are:

Control: No/poor scope verification. No/poor reviewing of the accomplishments and results of the project at selected project milestones. No/poor formalizing of project's deliverables acceptance.

Control: No/poor tracking of the project planning parameters against the project plan. No/poor periodic reviewing of the project's accomplishments and results. No/poor monitoring of the actual values of the project planning parameters against the project plan.

Control: No/poor quality control. No/poor monitoring of specific project results to determine whether they comply with the relevant quality standards and identifying ways to eliminate causes of unsatisfactory performance.

Control: Lack/poor of risk management. No/poor tracking of identified risks, monitoring residual risks, identifying new risks, executing risk response plans, and evaluating their effectiveness throughout the project life cycle.

Control: Lack poor contract administration. No/poor managing of the contract and relationship between buyer and seller: (1) reviewing and documenting how seller is performing or has performed to establish required corrective actions and provide a basis for future relationships with the seller; (2) managing contract related changes and, when appropriate, managing contractual relationship with the outside buyer of the project.

Control: Lack of frequent and efficient coordination. No/poor tracking of team member performance, providing feedback, resolving issues, and coordinating changes to enhance project performance.

Control: Team management. No/poor guarantee that the appropriate personnel are being assigned to be team members based on required knowledge and skills.
Control: Stakeholder management. Issues are not or are poorly resolved with relevant stakeholders.

A.4.1 Communication

Communication includes all the issues that prevent an effective exchange of information.

Communication: Uncertain team and location. The communication agents are not quite well known or, if they are known, it is not clear where their location is.
Communication: Ineffective team meetings. Meetings are neither well prepared nor conducted.
Communication: Ambiguity or multiple understandings. Ambiguity or multiple understandings due to different backgrounds, native languages, or even the simple fact that the same word might have different meanings even among people within the same culture.
Communication: Uncontrolled broadcasting of information. Information is broadcasted without control leading to security problems and information flood.
Communication: Lack or lack of strict enforcement of reading/replying rules. No/poor feedback. One is never sure if his messages have been received and acknowledged.
Communication: Leadership: executives' communication is thin and sparse. Inefficient high level communications.

A.4.2 Development execution

The development execution includes all the issues to the effective engineering, its subcategories are: requirements development, technical solution & integration, and verification & validation.

Requirements: Incomplete or incorrect picture of customer needs. Stakeholder needs, expectations, constraints, and interfaces were not transformed into customer requirements because: (1) they were not identified and collected; (2) their understanding was not developed with the requirements providers; or (3) their set was not ensured to necessary and sufficient.
Requirements: Conflicting requirements not solved. The requirements were not analyzed to balance stakeholder needs and constraints. Inconsistencies between the project plans and work products and the requirements were not identified.
Requirements: Retaining legacy requirements. Legacy requirements are carried on to new developments regardless their real need.

Requirements: No/poor requirements management. Changes to the requirements are not well managed as they pile during the project creating conflicts and inconsistencies.

Requirements: Poor translation of requirements in specs. Generated specs do not translate properly requirements: (1) established and maintained product and product-component requirements are not completely aligned to the customer requirements; (2) there are product components not related to requirements (inclusion of features not required by the customers).

Tech Solution: Bad exploration of solution space. Poor selection of the product-component solutions to satisfy the customer needs due to a bad development of alternative solutions, or to the use of a bad alternative selection criteria.

Strategy: Concept development is constrained. High reuse targets constrain the exploration of new concepts, limiting the development to incremental innovation.

Tech Solution: Bad integration planning. No/poor established and maintained procedures and criteria for integrating the product components. Bad definition of integration sequence.

Tech Solution: Complex product architecture with excessive interfaces. The architecture chosen is more complex than necessary (there are other simpler and as-good-as alternative). Complex product-component interfaces in terms of established and maintained criteria.

Tech Solution: Wrong level of modularity. Defined modules do not provide the minimum coupling level between them.

Tech Solution: Requirements are overlooked or not considered. Even though the analysis was well made and there is a good set of requirements, the technical solution team overlooks them.

Tech Solution: Lack of DFX. No use of Design for Excellence guidelines. Examples of design for excellence are: DFMA—Design for Manufacturing and Assembly; DTC—Design to Cost, DFE—Design for Environment. The choice of the X on the DFX depends on the dimension that the enterprise decides to optimize.

Tech Solution: Poor make, buy, or reuse analysis. No/poor/low evaluation whether the product components should be developed, purchased, or reused based on established criteria.

Tech Solution: Lack of concurrent engineering. No/poor concurrent engineering.

Tech Solution: Incremental PD. Making use of the architecture from previous models, when it may not fit well on the actual product to be developed (bad reuse).

Tech Solution: Low flexibility of technology. Chosen technology does not allow changes in product architecture or even changes in requirements.

Strategy: Technology readiness determined by demos under controlled environments. Technology readiness is wishful thinking, since no robust development is made (e.g.., design of experiments, failure mode analysis, etc.).

Verification & Validation: Bad testing. The planned testing (procedures, techniques, and tools) do not ensure the resulting product will perform as intended in the user's environment.

Verification & Validation: Premature validation. Premature confirmation (by examination and provision of objective evidence) that the particular requirements for a specific intended use are fulfilled.

Verification & Validation: Late verification. Late confirmation (by examination and provision of objective evidence) that specified requirements have been fulfilled.

Verification & Validation: Inappropriate environment. There is no/poorly established and maintained environment to support verification.

A.5 Resources

This category considers the issues related to the people, tools, and standards involved during the development.

A.5.1 People

The people are those who execute the development itself; they must have the proper knowledge, experience, and several other skills to positively contribute to the product development success.

People: Lack of knowledge. A person has basic, technical, or managerial knowledge/qualification below that needed to perform his/her role.

People: Lack of experience (know-how). A person has experience below that demanded by the project position.

People: Lack of confidence. Always need somebody else's opinion to take a decision.

People: Lack of critical thinking. Low critical faculty of a person that may cause: (1) excessive perfectionism resulting in over design; (2) difficulties to make deductions or inferences (creation of knowledge from previous knowledge); (3) no open-mindedness toward new solutions (low flexibility); (4) inability to see the whole (holistic view); (5) overconfidence; or (6) doing without thinking of consequences, impulsive.

People: Low commitment/motivation. Low commitment to the organization and to the work being performed. Person does not take the necessary care while performing his activities; includes also the "not my job" attitude and lack of motivation of a person.

People: Low discipline. No self-accountability of a person. Need to be supervised closely. Does not follow the process/plan. Does the job by "his way," even though there is an expected way.

People: Low communication skills. Poor communication skills of a person. Doesn't understand. Doesn't give feedback.

People: Team work issues. No team capability of a person. Does not work well in teams. Disaggregates the team, doesn't share knowledge etc.

People: Prejudices of a person. Person tries to protect his value by not sharing expertise.

People: Low leadership skills. A person has leadership skills below that demanded by the project position.

A.5.2 Tools

The tools are used by the people to perform their development tasks; they not only must be adequate for each task individually, but they also must be integrated at some level between themselves, allowing a smooth development flow.

Tool: Inadequate/obsolete. Insufficient investment to keep the tools up to date. Frequent tool producer version-ups also make it difficult to keep them up to date.

Tool: Low availability. Inadequate maintenance.

Tool: Lack of integrated solutions that meet the requirements of all users. Many partial solutions with compatibility issues. Set of tools allow the existence of "holes" (parts of the process that have no supporting tool or technique).

Tool: Complex equipment, tool or technique. The complexity may lead to misuse or simply prevent the use.

Tool: Low capacity. Tool capacity is below the necessary requirement.

Tool: No available support with problems. Tool/technique support for how to use is not available when needed.

Tool: Capable tools not fully used. Existing tools and applications are not known by potential users.

A.5.3 Standards

The standards are guidance to the work. Good standards, on the one hand, help reduce the variability of the development process, increasing the quality of each task outcome and the development success as a whole. Bad standards, on the other hand, provide misguidance and confusion by either requesting the wrong deliverables (do the wrong thing), or by suggesting a non-coherent or badly defined set of processes (do the thing wrong).

Standards: No, incomplete, or impractical standards on what/when to do something. The inexistence, incompleteness, or impracticability of the standards on what and when to do something prevent its use—people end up doing what they think is better when they want to do it. They might be impractical because: (1) they are too high level, providing no guidance; (2) the activities are badly sequenced, not optimizing the release of the resources and/or deliverables; (3) the set of activities within the process is complex or badly documented, impairing a clear picture of the whole; (4) there are superfluous activities; (5) etc.

Standards: No how-to standards. No guidelines to repeat previous success. Activities' description are badly written, lacking guidelines of how to perform.

Standards: Do not deliver or do not *quite* **deliver the right deliverables**. The standardized deliverables set does not match the real development needs (a deliverable must be useful to one's own or somebody else's job). As results: (1) deliverables are made just to stick to the process but are not used later; or (2) deliverables are too raw when first delivered, requiring additional work to be useful.

Standards: Feedback loop from distant (in time) activities. The right inputs to an activity may be available only very later in the process. The greater the distance between those activities the greater the rework if the assumptions previously made are proven incorrect.

Standards: Unclear/absent task ownership. The standard process does not clearly define all task owners.

Standards: Inadequate (too long) development time. The standard process is longer than necessary or does not allow its adaptation (inflexible) to faster development.

Standards: Inadequate (too short) development time. The development process creates too short and unrealistic development times.

Standards: Lack/poor contracting procedures. Lack/poor contracting standards, guidelines, and templates. No/ poor contracting knowledge management.

Standards: No/poor communication standards. There are no standards for the methods and medias to be used. Lack of standard terminology leading to ambiguity/accuracy. No or poorly designed templates for documentation.

References

1. Andrews K (1980) The concept of corporate strategy, 2nd edn. Dow-Jones Irwin
2. Steven JO (1989) The organizational culture perspective. Dorsey Press, Chicago
3. Ward A (2007) Lean product and process development. The Lean enterprise Institute, Cambridge, MA
4. Project Management Institute, PMI (2013) A Guide to project management body of knowledge (PMBOK® Guide), 5th edn. Project Management Institute, Newton Square

Appendix B
Commonly Used Program Metrics[2]

Requirements and specifications

- Number of customer needs identified
- Number of discrete requirements identified (overall system and by subsystem)
- Number of requirements/specification changes (cumulative or per unit of time)
- Requirements creep (new requirements/total number of requirements)
- Requirements change rate (requirements changes accepted/total number of requirements)
- Percent of requirement deficiencies at qualification testing
- Number of to-be-determined (TBD) requirements/total requirements
- Verification percentage (number of requirements verified/total number of requirements)

Electrical design

- Number of design review changes/total terminations or connections
- Number of post-design release changes/total terminations or connections
- Percent of fault coverage or number of faults detectable/total number of possible faults
- Percent of fault isolation
- Percent of hand assembled parts
- Transistors or gates designed per engineering man-month
- Number of prototype iterations
- First silicon success rate

Mechanical design

- Number of in-process design changes/number of parts
- Number of design review deficiencies/number of parts
- Number of drafting errors/number of sheets or number of print changes/total print features
- Drawing growth (unplanned drawings/total planned drawings)
- Producibility rating or assembly efficiency
- Number of prototype iterations
- Percent of parts modeled in solids

---

[2]Crow K (2001) Product Development Metrics. DMR Associates

© Springer International Publishing AG 2017
M.V.P. Pessôa and L.G. Trabasso, *The Lean Product Design
and Development Journey*, DOI 10.1007/978-3-319-46792-4

Software Engineering

- Man-hours per 1000 software lines of code (KSLOC)
- Man-hours per function point
- Software problem reports (SPR's) before release per 1000 software lines of code (KSLOC)
- SPR's after release per KSLOC
- Design review errors per KSLOC
- Code review errors per KSLOC
- Number of software defects per week
- SPR fix response time

Product assurance

- Actual mean time between failures (MTBF)/predicted MTBF
- Percent of build-to-packages released without errors
- Percent of testable requirements
- Process capability (Cp or Cpk)
- Product yield
- Field failure rate
- Design review cycle time
- Open action items
- System availability
- Percent of parts with no engineering change orders

Parts procurement

- Number of suppliers
- Parts per supplier (number of parts/number of suppliers)
- Percent of standard or preferred parts
- Percent of certified suppliers
- Percent of suppliers engaged in collaborative design

Enterprise

- Breakeven time or time-to-profitability
- Development cycle time trend (normalized to program complexity)
- Current year percent of revenue from products developed in the last "X" years (where "X" is typically the normal development cycle time or the average product life cycle period)
- Percent of products capturing 50 % or more of the market
- Percent of R&D expense as a percent of revenue
- Average engineering change cycle time
- Proposal win rate
- Total patents filed/pending/awarded per year
- R&D headcount and percent increase/decrease in R&D headcount

Portfolio and pipeline

- Number of approved projects ongoing
- Development work-in-progress (the non-recurring, cumulative investment in approved development projects including internal labor and overhead and external development expenditures and capital investment, e.g., tooling, prototypes, etc.)
- Development turnover (annual sales divided by annual average development work-in-progress)
- Pipeline throughput rate
- New products completed/released to production last 12 months
- Cancelled projects and/or wasted spending last 12 months
- Percent R&D resources/investment devoted to new products (versus total of new products plus sustaining and administrative)
- Portfolio balance by project/development type (percent of each type of project: new platform/new market, new product, product upgrade, etc.)
- Percent of projects approved at each gate review
- Number of ideas/proposed products in the pipeline or the investigation stage (prior to formal approval)

Organization/Team

- Balanced team scorecard
- Percent of project personnel receiving team building/team launch training/facilitation
- Average training hours per person per year or percent of payroll cost for training annually
- IPT/PDT turnover rate or average IPT/PDT turnover rate
- Percent of core team members physically collocated
- Staffing ratios (ratio of each discipline's headcount on project to number of design engineers)

Program Management

- Actual staffing (hours or headcount) versus plan
- Personnel turnover rate
- Percent of milestone dates met
- Schedule performance
- Personnel ratios
- Cost performance
- Milestone or task completion versus plan
- On-schedule task start rate
- Phase cycle time versus plan
- Time-to-market or time-to-volume

Product

- Unit production cost/target cost
- Labor hours or labor hours/target labor hours
- Material cost or material cost/target material cost
- Product performance or product performance/target product performance or technical performance measures (e.g., power output, mileage, weight, power consumption, mileage, range, payload, sensitivity, noise, CPU frequency, etc.)
- Mean time between failures (MTBF)
- Mean time to repair (MTTR)
- System availability
- Number of parts or number of parts/number of parts for last generation product
- Defects per million opportunities or per unit
- Production yield
- Field failure rates or failure rates per unit of time or hours of operation
- Engineering changes after release by time period
- Design/build/test iterations
- Production ramp-up time
- Product ship date versus announced ship date or planned ship date
- Product general availability (GA) date vs. announced/planned GA date
- Percent of parts or part characteristics analyzed/simulated
- Net present value of cash outflows for development and commercialization and the inflows from sales
- Breakeven time (see above)
- Expected commercial value
- Percent of parts that can be recycled
- Percent of parts used in multiple products
- Average number of components per product

Technology

- Percent of team members with full access to product data and product models
- CAD workstation ratio (CAD workstations/number of team members)
- Analysis/simulation intensity (analysis/simulation runs per model)
- Percent of team members with video-conferencing/desktop collaboration access/tools

Glossary

5 Whys Toyota's practice of asking why five times to solve problems at the root cause

A3 Process The A3 process is a Toyota-pioneered practice of getting a problem, an analysis, a corrective action, and an action plan written down on a single sheet of large paper, often with the use of simple graphics

Ambiguity The existence of multiple conflicting interpretations of the information held and required which leads to a lack of consistent information

Analysis A check action through evaluation equations, graphs, data reduction, extrapolation of results, or reasoned technical argument, that specified requirements for a material or service have been met

Benefits Results of the deliverables used by the stakeholders which may be tangible or intangible

Calculation Performing mathematical or computer simulations

Chief engineer A prestigious position in the company, accountable for project's results

Chief engineer staff A team that responds directly to the chief engineer and shares with him/her the responsibilities of representing the voice of the customer and giving common vision/goal to all functional teams involved in the PD

Competitor teardown and analysis An exercise that provides an opportunity to learn about competitors

Continuous improvement and adaptation The ability to move toward a new desired state through an unclear and unpredictable territory by being sensitive to and responding to actual conditions on the ground

© Springer International Publishing AG 2017
M.V.P. Pessôa and L.G. Trabasso, *The Lean Product Design and Development Journey*, DOI 10.1007/978-3-319-46792-4

Corporate strategy The pattern of decisions in a company that determines and reveals its objectives, purposes, or goals; produces the principal policies and plans for achieving those goals; and defines the range of business the company is to pursue, the kind of economic and human organization it is or intends to be, and the nature of the economic and non-economic contribution it intends to make to its shareholders, employees, customers, and communities.

Cross-checking One method to discover problems and check quality by requiring several groups to check the same parts/data independently

Daily wrap-up meetings Meetings held at the end of each day, typically on the shop floor where the work is being done, attended by all key participants, including suppliers. They clarify assignments, and generally aid in real-time, course-correction decisions

Deliverables The outputs from the product lifecycle (which includes the PDP). A deliverable is tangible and has one or more specific recipients

Demonstration The display of features, performance, and operational capacity of an item, equipment, or system where success is found only through behavioral observation and/or results. Tests that require a simple quantitative verification measure, such as weight, size, or time, to perform tasks are included in this category

Design patterns In software engineering, a design pattern is a general reusable solution to a commonly occurring problem within a given context in software design

Dominant design The design that wins the allegiance of the marketplace and that competitors and innovators must adhere, making many of the performance requirements implicit in the design itself

Earned value analysis (EVA) A project management methodology that integrates scope, schedule, and resources. The EVA objectively measures the cost and schedule performance and progress of the project by comparing costs (actual and planned) and value

Eternal stakeholders The ones who pull value from the product development program's final results (the product and/or services). They can be encountered when we consider the "Product/Process Follow-up" and the "Product Discontinuation" process groups in the product development process

Functional program teams The functional divisions related to the development program itself

Gemba (現場) The actual place where the real work is done or where the problems and issues are arising

Genchi genbutsu (現地現物) "Go and see for yourself." Going to the place where the actual working is being performed, or where the problems and issues are arising and seeing for yourself, and not losing any details from the real thing

Guest engineers Engineers from suppliers who reside full time in the LPDO product development office

Hansei (反省) Self-reflecting, identifying things that did not go well, and then taking responsibility

Hourensou (ほうれん草) You must frequently report the progress of your work and its result (*houkoku*, 報告/report), you must pass on the actual information without your opinion (*renraku*, 連絡/communication), and you must ask for an advice from a peer, a mentor, or a leader when you can't decide (*soudan*, 相談/ discussion or ask for an advice)

Ijiwaru The practice of testing subsystems to the point of failure

Inspection An action of observation, visual examination or investigation against relevant document to confirm the compliance of the material or system with the technical requirements

Integrated product development A PD approach where the requirements from the areas constitutive of the product lifecycle such as design, manufacturing, assembly, maintenance, disposal etc. are considered, weighed, discussed, and balanced at the conceptual phase of the PDP

Integrative design variable Represents a design variable with a specified target value, which affects and is affected by most of the design decisions

Internal stakeholders The ones the pull value from the product development program's intermediate results. They can be encountered when we consider the "Design & Development" and the "Production/Ramp-up" process groups in the product development process

Jidoka (自働化) Loosely translated as "automation with a human touch," meaning that when a problem occurs, the equipment should be stopped immediately, preventing defective products from being produced

Just-in-Time (ジャストインタイム) Making "only what is needed, when it is needed, and in the amount needed."

Kaizen (改善) A continuous improvement strategy where employees work together proactively to achieve regular incremental improvements to a process

Kaikaku (改革) A more fundamental, larger scale, and radical process change than the kaizen

Kanban (看板, **also** かんばん) A control card used to control the production flow. When a process refers to a preceding process to retrieve parts, it uses a *kanban* to communicate which parts have been used

Kentou (**Study Phase**) The early phase of the PDP, in which the objective is solving problems and resolving conflict, thus addressing the roots of variation, and segregating it from the rest of the PDP. During the study phase the product is conceived, a performance envelope is defined, and the solution space is explored in order to find a balanced (value/risk) design

Keyretsu A supplier model where both companies hold equity in each other

Lean wheel system A pictorial model that shows the elements that support the Lean Product Development and their relationship

Lean thinking (lean philosophy) A way to specify value, line up value-creating actions in the best sequence, conduct these activities without interruption whenever someone requests them, and perform them more and more effectively. Womack and Jones (2003, p. 15)

Market life cycle All stages from the product conception until its discontinuity, while the enterprise works to make and keep the product competitive

Module development teams (MDT) Cross-functional teams responsible for each product subsystem

Nemawashi (根回し) Laying the groundwork, building consensus, literally: "going around the roots." From the original meaning of digging around the roots of a tree, to prepare it for a transplant, in business nemawashi brings the same careful preparation before attacking a problem

Organizational culture The sum of all the written and unwritten rules, attitudes, behaviors, beliefs, and traditions, which contribute to the unique social and psychological environment of an organization

Overburden (Muri, 無理) Pushing a machine, process, or person beyond natural limits

Obeya or *Oobeya*(大部屋) A large room, war room

Product platform A collection of the common elements, especially the underlying core technology, implemented across a family of products

Product All the results from the PDP, not limited to physical products, but also encompassing services, product-as-service, and even complete value chains, which are aimed to fulfill the customer and user needs

Product development process The set of activities beginning with the perception of a market opportunity aligned with the company's competitive strategy and technical capacity, and ending in the production, sale, and delivery of a product, while considering all aspects that will turn and keep the product competitive in the market until its discontinuity

Product use analysis Understanding how the value expected meshes with the program's vehicle performance and characteristics; providing deep understanding of the customer experience/expectations with/from the product

Pull event is a verification and/or validation event that pulls the necessary information from the development team at the right moment during the PD project, the pull events set guarantee the value flow, make quality problems visible, and create knowledge

Resident engineers Engineers exchanged on temporary assignment both within Toyota and with affiliated companies

Serial product development A PD approach that only takes into account the functionality of the product during the conception phase of the PDP

Simultaneous engineers Key production engineers assigned to MDTs and that function as full-time representatives of their manufacturing disciplines

Stakeholders Individuals or organizations actively involved in development, or whose interests may be affected by its execution or completion (either successful or failure)

Supply chain A downstream flow of goods and supplies from the source to the customer

Team A group of people that has the following four characteristics: they share a mission, are committed to this commitment, have complementary capacity, and ephemerality (the team ceases after finishing the mission)

Test The verification of action through the full exercise of the item, equipment, or system under appropriate controlled conditions in accordance with approved test procedures. The test can be a subsystem (T1) and the integrated product (T2)

Trade-off curves A subsystem's performance on one characteristic is mapped on the X-axis while the other is mapped on the Y-axis and a curve is then plotted to illustrate subsystem performance relative to the two characteristics

Uncertainty The knowledge gap between the supposed and the verified characteristics which lasts while the development is not completed

Unevenness (*Mura*, 斑 **or** ム ラ) An irregular production schedule or fluctuating product volumes caused by internal problems

Value For a given stakeholder, value is the total and balanced perception, resulting from the various benefits delivered through the product/process lifecycle

Value chain Activities for a firm operating in a specific industry. It models of how businesses receive raw materials as input, add value to the raw materials through various processes, and sell finished products to customers. Therefore, it comprises all the organization's primary and support activities, not forgetting all the interfacing activities with other organizations within the supply chain

Value engineering (VE) An organized/systematic approach that analyzes the functions of systems, equipment, facilities, services, and supplies to ensure they achieve their essential functions at the lowest life-cycle cost consistent with required performance, reliability, quality, and safety

Value Function Deployment (VFD) is a technique adapted from the Quality Function Deployment (QFD) that applies the lean principles based on value creation and waste reduction to derive a project activity network that entails a sequenced set of confirmation events

Value targeting process An understanding of what each stakeholder values; it provides a deep knowledge of their needs, particularly of the customers

Waste (*Muda*, 無駄 **or** ムダ) All elements of a process that only increase cost without adding value or any human activity that absorbs resources but creates no value; any activities that lengthen lead times and add extra cost to the product for which the customer is unwilling to pay

Index

© Springer International Publishing AG 2017
M.V.P. Pessôa and L.G. Trabasso, *The Lean Product Design and Development Journey*, DOI 10.1007/978-3-319-46792-4

Printed in the United States
By Bookmasters